Elizabeth Connor, MLS, AHIP

Internet Guide
to Food Safety and Security

T0186780

Pre-publication
REVIEWS,
COMMENTARIES,
EVALUATIONS . . .

"The *Internet Guide to Food Safety and Security* by Elizabeth Connor has been written for the English-language professional and consumer health audiences. The material covered emphasizes Internet sites within North America but also includes some links from the United Kingdom, other European countries, and Australia. Connor has authored and edited a number of articles and books dealing with the Internet and electronic resources.

This work is a comprehensive compilation of annotated links for those 'interested in staying healthy when handling, preparing, and storing food at home and eating foods prepared by others.' With the emergence of the global economy, there is a need to 'determine the source, safety, or security of food available for consumption.'

In the introduction, the author notes that consumers of health information should be extremely careful about medical information or advice that is ob-
tained through the Internet. Individuals should closely check the currentness, accuracy, and source of information of the online resource. The author notes the need for criteria in evaluating Internet information.

The book is a series of well-organized, category-based lists of Internet information. These topical or format categories range from comprehensive sites to specialized sites of interest such as governmental medicine. Each link includes an annotation that notes the authority of the resource, a summary of its scope, and highlights key features. A check mark is used to denote major resources with authoritative and original content.

Besides the subject categories of links, the work contains an extensive list of useful search engines and directories, a detailed explanation of Web site addresses, and a comprehensive glossary derived from numerous online sources."

Lenny Rhine, PhD
University Librarian,
Health Science Center Libraries,
University of Florida

More pre-publication
REVIEWS, COMMENTARIES, EVALUATIONS . . .

"Connor has written a thorough and informative book dealing with many different aspects of food safety and security on the Internet. *Internet Guide to Food Safety and Security* should prove to be useful for the consumer, public health student, or librarian. The introductory chapter explains how to evaluate a Web site for authoritative and peer-reviewed sources, and lists the Web addresses for various search engines that can be used to locate these sources. The chapters dealing with specific safety issues and concerns, disease, and legislation cover a wide variety of public health issues. This book will be a welcome addition to any reference shelf."

Kathy D. Broyles, MLS, AHIP
Public Services Librarian,
University of North Texas
Health Science Center

"This gold mine of information provides annotations and addresses for hundreds of Web sites related to food safety and security and related concerns. This book is a worthwhile addition for public libraries, consumer health collections, and health sciences libraries. Academic libraries supporting programs in agriculture, food sci-

ence and service, nutrition, and public health will find this work to be particularly relevant.

Connor's concise annotations provide the reader with information needed to evaluate Web sites as well as the background information necessary to better understand the importance of the food safety issues addressed. Any librarian interested in creating electronic pathfinders or research guides related to food safety issues will find *Internet Guide to Food Safety and Security* to be a great time-saver because Connor has preselected the best sites. In addition, a checkbox symbol identifies sites with original and authoritative content so that the reader can quickly identify the pearls. A convenient glossary also provides needed explanation of unique terms associated with food safety as well as the language used to describe various types of Web tools.

Connor's style is approachable for general readers and students, but is also useful for faculty and librarians as well. I recommend reading this book at your computer, because you will feel the need to begin looking up many sites as you read. Once you begin carefully examining the sites, you might never see your favorite foods or restaurants in the same way."

David A. Nolfi, MLS, AHIP
Health Sciences Librarian,
Duquesne University

The Haworth Information Press®
An Imprint of The Haworth Press, Inc.
New York • London • Oxford

Internet Guide
to Food Safety and Security

THE HAWORTH INFORMATION PRESS®
Haworth Internet Medical Guides
M. Sandra Wood, MLS
Editor

The Guide to Complementary and Alternative Medicine on the Internet by Lillian R. Brazin

Internet Guide to Travel Health by Elizabeth Connor

Internet Guide to Cosmetic Surgery for Women by M. Sandra Wood

Internet Guide to Food Safety and Security by Elizabeth Connor

Internet Guide
to Food Safety and Security

Elizabeth Connor, MLS, AHIP

The Haworth Information Press®
An Imprint of The Haworth Press, Inc.
New York • London • Oxford

For more information on this book or to order, visit
http://www.haworthpress.com/store/product.asp?sku=5373

or call 1-800-HAWORTH (800-429-6784) in the United States and Canada
or (607) 722-5857 outside the United States and Canada

or contact orders@HaworthPress.com

Published by

The Haworth Information Press®, an imprint of The Haworth Press, Inc., 10 Alice Street, Binghamton, NY 13904-1580.

PUBLISHER'S NOTES
The development, preparation, and publication of this work have been undertaken with great care. However, the publisher, employees, editors, and agents of The Haworth Press are not responsible for any errors contained herein or for consequences that may ensue from the use of materials or information contained in this work. The opinions expressed by the author(s) are not necessarily those of The Haworth Press, Inc.

Due to the ever-changing nature of the Internet, Web site names and addresses, though verified to the best of the publisher's ability, should not be accepted as accurate without independent verification.

Cover design by Marylouise E. Doyle.

International Food Safety Icons used on the cover courtesy of the International Association for Food Protection.

Library of Congress Cataloging-in-Publication Data

Connor, Elizabeth, MLS.
    Internet guide to food safety and security / Elizabeth Connor.
        p. cm.
    Includes bibliographical references and index.
    ISBN-10: 0-7890-2631-7 (hc.)
    ISBN-10: 0-7890-2632-5 (pbk.)
    ISBN-13: 978-0-7890-2631-6 (hc. : alk. paper)
    ISBN-13: 978-0-7890-2632-3 (pbk. : alk. paper)
    1. Food adulteration and inspection—Computer network resources—Directories. 2. Food—Safety measures—Computer network resources—Directories. 3. Internet addresses—Directories. I. Title.

TX531.C613 2005
025.06'36319264—dc22

                                                        2005005953

To fearless and courageous librarians everywhere

ABOUT THE AUTHOR

Elizabeth Connor, MLS, AHIP, is Assistant Professor of Library Science and Science Liaison at the Daniel Library of the Citadel, the Military College of South Carolina in Charleston, and is a distinguished member of the Academy of Health Information Professionals. She has held library leadership positions at teaching hospitals and medical schools in Maryland, Saudi Arabia, Connecticut, South Carolina, and the Commonwealth of Dominica. She has authored several peer-reviewed articles about electronic resources, search engines, chat reference, and medical informatics, and has published more than 50 book reviews in *Library Journal, Against the Grain, Bulletin of the Medical Library Association, Journal of the Medical Library Association, Medical Reference Services Quarterly,* and the *Post & Courier.* She is the author of *Internet Guide to Travel Health* and the editor of *A Guide to Developing End User Education Programs in Medical Libraries.* She has served as the Associate Editor (International) for the *Bulletin of the Medical Library Association* and has edited the From the Literature column for *Medical Reference Services Quarterly.* Ms. Connor currently manages the book review process for *Medical Reference Services Quarterly* and is co-editor of *Journal of Electronic Resources in Medical Libraries.*

CONTENTS

Preface

Food is an essential part of daily existence. Human health and well-being are mutually dependent on animals and plants, and the safety and security of the environment as a whole. Regardless of whether you live to eat or eat to live, health consumers and health professionals alike seek authoritative, reliable, and up-to-date information about the source, production, storage, handling, and preparation of animal and plant foods. Persons who travel for business and pleasure are concerned with avoiding or reducing health risks associated with foodborne illnesses. Contemporary concerns and controversies include food irradiation, genetic engineering, food inspection, food dyes and additives, hormones and antibiotics administered to cattle and poultry, chemicals used in animal feeds, pesticides, bioterrorism, water quality, and other practices that affect the food cycle.

The food chain and food production cycle are intricate and complex processes. Humans depend on weather, growth, decomposition, and manufacturing processes that result in edible food and potable drink. According to the *2001 FDA Food Code*, <http://www.cfsan.fda.gov/~dms/fc01-toc.html>, "an estimated 76 million illnesses, 323,914 hospitalizations, and 5,194 deaths are attributable to foodborne illness in the United States each year. The estimated cost of foodborne illness is $10-$83 billion annually."

This comprehensive compilation of annotated links serves as a handy, useful, and easy-to-consult guide for herbivores, carnivores, and omnivores alike who are interested in staying healthy when handling, preparing, and storing food at home and eating foods prepared by others. The emphasis is on English-language information for North Americans, with some links from the United Kingdom, Australia, and some European countries.

Chapter 1

Introduction

Healthy eating involves planning, preparation, and awareness of situations and potential health risks. In our increasingly global economy, it has become more difficult to determine the source, safety, or security of foods available for consumption. People travel the world for a variety of reasons, including pleasure, work, and education. The case of traveling thousands of miles in less than a day's time exposes us to myriad health risks caused by different climate conditions, organisms, populations, flora and fauna, and food and drink.

Discerning individuals can use the Internet to find updated and authoritative information on a variety of topics, including consumer health. The Internet has the potential to improve health knowledge and to increase awareness of health risks. **The Pew Internet and American Life Project** <http://www.pewinternet.org/> reports that "fifty-two million American adults, or 55% of those with Internet access, have used the Web to get health or medical information."

ANATOMY OF A WEB SITE ADDRESS

Hypertext transfer protocol (http) is the set of standards used to represent content on the World Wide Web. Although many Web browsers no longer require the http:// prefix when entering site addresses, other prefixes are understood by browser software to connect to other types of Internet resources. For example, the telnet:// prefix is used to establish a Telnet connection, which allows remote logins to resources such as electronic catalogs. The ftp:// prefix uses file transfer protocol to transmit files from one computer to another. The gopher:// prefix is used to connect to gopher content, an Internet protocol and organizational structure that was developed before the World

Wide Web. Each Web site address is composed of distinct and meaningful parts that describe the host computer, directory, and file name:

<protocol://host.domain.suffix.suffix/directory>

For example, in the address <http://tti.tamu.edu/media/>, tti (Texas Transportation Institute) is the host name; tamu (Texas A&M University) is the domain; .edu indicates an educational institution; and media is the directory. In the address for MedlinePlus's information about food safety <http://www.nlm.nih.gov/medlineplus/foodsafety. html>, the domain is the National Library of Medicine (NLM) at the National Institutes of Health (NIH), a group of government agencies (.gov); the directory is MedlinePlus; and the file name is foodsafety.

Many Web users are familiar with the use of file extensions such as .htm, .html, and .pdf at the end of site addresses. Hypertext markup language (HTML) is used to create and display Web content, as indicated by .htm or .html at the end of file names, but this practice is seen less frequently because of changes in site design. Documents in Adobe's portable document format use the .pdf extension. Web designers that use "server-side includes" (SSI) use .shtml instead of .htm or .html. Dynamic files created with ColdFusion software may have .cfm at the end. Although .php originally meant personal home page, it now means hypertext preprocessor, signifying the use of server-side scripting language. Sites that use Java scripting software may show .jsp (Java server page) as part of the file name. If a site uses Microsoft scripting software instead of Java, the file suffix is .asp (active server page).

The ease with which Web sites are designed and content can be uploaded has resulted in many temporary, redesigned, or outdated sites. Dead links result when a site changes file names, alters the site navigation, or stops publishing. If a particular site address no longer functions, delete the /directory, /filename.htm, or /filename.html part of the address, and use the host.domain.suffix parts. After the site loads, use the site's search function to find the specific document or section needed. If this approach does not work, try the **wayback machine,** <http://www.archive.org/web/web.php>, which provides a simple interface for searching billions of archived Web pages dating back to 1996.

Web site addresses use a variety of organizational and geographic suffixes that are meaningful. Table 1.1 lists common address suffixes

TABLE 1.1. Site Type and Geographic Suffixes

Site Type Suffix	Geographic Suffix
.com—Commercial sites	.au—Australia
.edu—Educational sites	.ca—Canada
.gov—Government sites	.ch—Switzerland
.mil—Military sites	.ie—Ireland
.net—Commercial sites that provide network services	.nz—New Zealand
.org—Organization or association sites	.uk—United Kingdom

and their meanings. There is nothing inherently suspicious or problematic about scientific content featured on commercial Web sites, or sites with a .com suffix at the end of the address. Reputable educational institutions such as the Mayo Clinic and Johns Hopkins University, for example, maintain authoritative consumer health information on .com sites with content that is separate from their .edu sites. Some excellent consumer health sites sell brochures and other publications, but the sites featured in this guide provide freely available information, including some sites that require completion of a registration process to personalize future site interactions or association membership for full access to all content on the site.

EVALUATING WEB CONTENT

Consumers of health information should be particular and skeptical about medical information or advice obtained through the Internet. The currency, accuracy, and source of health-related information are very important factors to consider. Laypersons should be as discerning as health professionals when distinguishing between anecdotal information and content derived from authoritative and peer-reviewed sources.

Health on the Net (HON) Foundation <http://www.hon.ch/> is an organization based in Switzerland that developed **MedHunt,** an English/French medical search engine, and a set of standards for evaluating sites with medical content. The HON Code of Conduct rates Web sites according to whether a particular site with medical content

- explains qualifications for dispensing advice or developing content;
- maintains confidentiality when handling medical information;
- attributes information derived from other sources;
- indicates the date content was modified or revised;
- lists contact information for content developers;
- identifies sources of funding or sponsorship;
- explains the use of advertisements or sale of products; and/or
- distinguishes original informational content from promotional content.

USING SEARCH ENGINES/DIRECTORIES

Although Internet applications date back to 1969 and were originally developed to allow government agencies to communicate and share information with one another, the World Wide Web was not introduced until 1990. Gopher, the first Internet search tool, appeared in 1991, but the development of graphical browser tools (Mosaic, Netscape) and search engines/directories accelerated the growth, development, and acceptance of the World Wide Web.

The proliferation of search engines/directories has helped the Web evolve into a tool for daily living, but too often a typical search query yields thousands of marginally relevant results, with many dated or extinct links. Search engines have advantages and drawbacks, and it is worthwhile to learn the features of a few to serve a variety of needs.

A search engine delivers dynamically generated results based on the words typed into the search box. Search directories are somewhat static groupings of categorized sites and tend to be smaller in scope than search engines. For subjects related to food safety and security, it may be more productive to focus on several sites with reliable health content (United States Department of Agriculture, Food and Drug Administration, Centers for Disease Control and Prevention, National Library of Medicine's MedlinePlus, World Health Organization) than to enter keywords into search engine interfaces and spend hours sorting through links of dubious quality or authority.

Search engines/directories vary greatly in size and in how they are compiled, updated, and organized. **Search Engine Showdown** <http://www.searchengineshowdown.com/> and **Search Engine Watch** <http://searchenginewatch.com/> are excellent sources of information about how specific search tools work and the relative size, advantages, and features of each. The following search engines/ directories are useful for searching a variety of topics and were used to locate links described in this book:

- **Google** <http://www.google.com/>: Google is an excellent all-purpose resource for searching or browsing content, including publicly accessible Web sites, news group messages, images, and news information. The subject categories in the **Google Directory** <http://directory.google.com/> can be browsed and searched. Google's size is estimated at more than 3.3 billion text documents. Features include the caching of old pages and providing links to similar content.
- **All the Web** <http://www.alltheweb.com/>: At 3.2 billion pages, All the Web is considered one of the top search engines and is a close second to Google in terms of ease of use and comprehensiveness. Use this search engine if Google does not deliver the results expected. The search mechanism is powered by Yahoo!
- **Yahoo!** <http://www.yahoo.com/>: Yahoo! is the first Web search directory, and although it is small in size (approximately 1.7 million pages), its advantages include compilation by humans, organization, and ease of use. If Yahoo! exhausts its index, it refers the search query to its own search engine, introduced in February 2004. Yahoo! also uses Google to power its image search feature, and Yahoo! editorial and technological means to power its news search feature.
- **Teoma** <http://www.teoma.com/>: Although Teoma is smaller than other top-rated engines at 1.5 billion pages indexed, it proves its worth by delivering relevant results. Owned by **Ask Jeeves** <http://www.ask.com/>, this crawler-based engine has useful features, including the clustering of results retrieved from the same site and linking to more results.
- **Vivisimo** <http://vivisimo.com/>: Vivisimo is a meta-search engine with a useful clustering mechanism. Meta-search engines search several search engines at once. A quick search on food

safety yields a group of 170 sites clustered into different folders: University (21), National Food Safety (9), Center for Food Safety (10), Food Safety Education (9), Food Safety Information (8), Kids (13), Food Safety and Quality (10), and so forth.

Several search engines and directories focus on specific areas of interest such as government or medicine. These specialized resources may be useful to research food safety and security topics:

- **FirstGov.gov** <http://www.firstgov.gov/>: FirstGov is the U.S. government's Web portal, an aggregated interface to content available on federal and state government sites. **FirstGov for Science** <http://www.science.gov/> is a portal subset that focuses on science, specifically authoritative information available from twelve U.S. government agencies, including the Departments of Agriculture, Commerce, Defense, Energy, Education, Health and Human Services, Interior, the Environmental Protection Agency, NASA, the National Science Foundation, and the Government Printing Office. In addition to searching these agencies and their databases, it is possible to browse topic areas such as Agriculture & Food; Applied Science & Technology; Astronomy & Space; Biology & Nature; Computers & Communication; Earth & Ocean Sciences; Energy & Energy Conservation; Environment & Environmental Quality; Health & Medicine; Math, Physics, & Chemistry; Natural Resources & Conservation; and Science Education (see Figure 1.1).
- **MedHunt** <http://www.hon.ch/medhunt/>: HON is known for its approval system for health-related sites. HON also maintains this medical search engine which annotates and ranks search results.
- **<http://searchedu.com> <http://searchmil.com> <http://search gov.com>:** Despite the .com suffix on each of these site addresses, the interfaces retrieve results that are limited to education (.edu), military (.mil), and government (.gov) sites, respectively.
- **Scirus** <http://www.scirus.com/>: As of January 2004, Scirus "covers over 170 million science-related Web pages." Scirus focuses on scientific, peer-reviewed journal content including e-prints, preprints, citations, and full-text literature. Nonscientific sites are filtered out of the search results. Results can be restricted

to specific years, journals, authors, and subjects within science, technology, and medicine.

The Invisible Web refers to content that is not easily accessible by normal search engines because of the way some information is organized and how search engines find links. To tease out valuable content featured in deeper layers of sites or within some Web-based databases, try **Search Adobe PDF Online** <http://searchpdf.adobe.com>, **InfoMine** <http://infomine.ucr.edu/>, **ProFusion** <http://www.profusion.com/>, or **Spire Project Light** <http://spireproject.com/spir.htm>.

STAYING HEALTHY AND WELL-INFORMED

The world's food supply is vulnerable throughout its production, storage, and preparation stages, and can be threatened by a variety of dangers from naturally occurring organisms to contamination to terrorist threats. Consumers can improve their knowledge related to food safety and security by focusing on authoritative sources of information and avoiding unsubstantiated consumer health information.

Healthy People 2010 <http://www.healthypeople.gov/> is an effort organized by the Office of Disease Prevention and Health Promotion and other U.S. federal agencies to encourage American citizens to take personal responsibility for their health by developing effective food safety habits, reducing foodborne pathogens, and employing other strategies. Now more than at any other time in history, health knowledge, preventive measures, and treatments are available to improve and extend the quality of life for all.

American society thrives on the dissemination of urban legends, meaning that recent but largely untrue stories are spread by word of mouth (and lately, by e-mail and Web sites) and reach epic proportions. Fictitious examples related to food safety and security include the presumed use of mutant chicken parts by a major fast food company and the presence of worms in hamburger meat. To help determine whether an anecdote is truth or folklore, consult **Urban Legends Reference Pages** <http://www.snopes.com/>.

The United States maintains two national libraries that provide services and resources in support of food safety and security and other

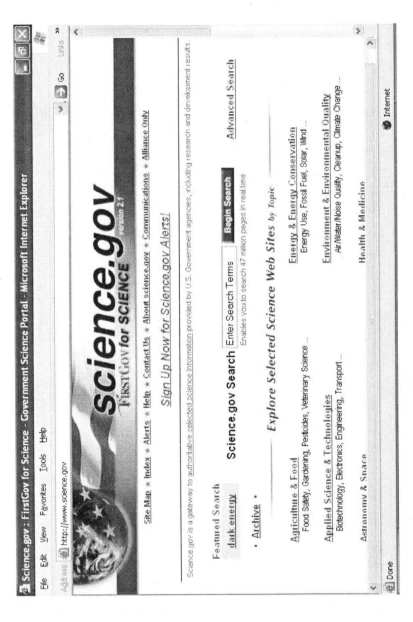

FIGURE 1.1. FirstGov for Science <http://www.science.gov/>

scientific subjects. The **National Agricultural Library** (NAL) <http://www.nal.usda.gov/> in Beltsville, Maryland, strives to improve life by providing access to agricultural information, most notably through its collections, services, and resources such as AGRI COLA, a bibliographic database that covers many aspects of agriculture, forestry, and animal science. The NLM <http://www.nlm.nih.gov/> in Bethesda, Maryland, is the largest biomedical library in the world and produces MEDLINE, MedlinePlus, and other bibliographic and full-text databases.

Various government sites and universities with well-known agricultural studies programs such as Cornell, Clemson, Iowa State, Texas A&M, Purdue, Rutgers, and others are excellent sources of information about topics related to the safety and security of food supplies, and safe handling, preparation, and storage of raw and cooked foods.

Although this guide is intended to be as thorough as possible, use the sites marked with the ☑ symbol to save considerable time and effort when researching the safety and security of food. Consult the glossary to learn the definitions of words that are unfamiliar. This focus on the safety and security of food is intended to educate and improve awareness, not to disturb or alarm. Scientific knowledge increases exponentially, and foods that were safe a few years ago can be dangerous today because of changes in economics, environment, agriculture, manufacturing, and society. Wash your hands, understand the ingredients and contaminants in food and drink, and stay healthy by being vigilant and informed. May you thrive and prosper in this wonderful and ever-changing world.

BIBLIOGRAPHY

Connor, E. *Internet Guide to Travel Health.* Binghamton, NY: The Haworth Press, 2004.

Chapter 2

General Food, Health, and Agriculture Sites

The sites included in this section cover many aspects of food safety and security, and are excellent resources for exploring various topics related to staying healthy while preparing, eating, and storing foods. Consult the more specific chapters in this book to focus on individual issues, concerns, or controversies related to this subject. The symbol ☑ is used to denote major resources with authoritative and original content.

AgriBiz
<http://www.agribiz.com/>

AgriBiz is a consulting firm that specializes in agriculture. This site focuses on agricultural business news, research articles, and market and analysis information. News sources include daily food business headlines, *Agricultural Biotechnology and Food Safety Daily News, Supply Chain Management,* and the U.S. Department of Agriculture's Annual Agricultural Outlook.

Agriculture Network Information Center (AgNIC)
<http://Paurel.nal.usda.gov:8080/agnic/>

Hosted by the U.S. Department of Agriculture (USDA), AgNIC combines agricultural information and subject specialists from the National Agricultural Library (NAL), land-grant universities, and various agricultural organizations, as well as related citizen groups and government agencies. Subject areas include economics, animal science, food science, plant science, forestry, and natural resources. Site features include a calendar of events, news, discussion forum, and browsing by topic.

AgriFood
<http://www.agrifood.com/>

AgriFood focuses on three industry sectors: agriculture, food, and biotechnology. The company provides consulting services and think tank position papers and research reports. Site features include news; think pieces; position papers; events; discussion board; and links to agroterrorism, agriculture, biotechnology, and food resources.

AGRIGATE
<http://www.agrigate.edu.au/>

AGRIGATE describes itself as an agriculture information gateway for Australian researchers, and is a project of the libraries of the Universities of Melbourne, Adelaide, and Queensland and the Commonwealth Scientific and Industrial Research Organisation (CSIRO). Site subject areas include agribusiness, farming systems, aquaculture, field crops, food sciences, meteorology, and viticulture, and the site can be browsed or searched.

AgriGator: Agricultural and Related Information
<http://agrigator.ifas.ufl.edu/ag.htm>

The Institute of Food and Agriculture Sciences (IFAS) at the University of Florida has compiled this collection of agricultural and biological information resources. The site also features AgriForum, a Web-based program to facilitate the discussion of agricultural topics, agricultural maps, and agricultural sites organized into categories including state, government agency, publications, conferences, and associations.

☑ Centers for Disease Control and Prevention (CDC)
<http://www.cdc.gov/>

The CDC is a U.S. federal agency located in Atlanta, Georgia, that focuses on protecting health and safety by monitoring and preventing disease. Health promotion and education activities relate to eating nutritious foods, being physically active, and avoiding tobacco use. Health agencies report diseases to the CDC in an effort to understand disease causes and outbreaks. CDC publications related in part to food safety and security include *Morbidity and Mortality Weekly Report* (MMWR), *CDC Fact Book 2000/2001*, and others.

Cooperative State Research, Education, and Extension Service (CSREES)
<http://www.csrees.usda.gov/>

CSREES, of the USDA, concerns itself with advancing "knowledge for agriculture, the environment, human health and well being, and communities." Site features include information about state extension offices, funding opportunities, news, workshops, and more.

FirstGov for Consumers
<http://www.consumer.gov/>

This site of aggregated U.S. government information for consumers focuses on several areas of concern: food, product safety, health, home and community, money, transportation, children, careers and education, and technology, as well as other topics such as identity theft and outdoor recreation. Site features related to food safety and security include fruits and vegetables, labeling, meat and poultry, nutrition, product recalls, safety, seafood, and topics in the spotlight.

☑ Food and Drug Administration (FDA)
<http://www.fda.gov/>

The U.S. Food and Drug Administration (FDA) is part of the Department of Health and Human Services and is responsible for "protecting consumers and promoting public health." Site features include news, hot topics, activities, newsletters sent by e-mail, and publications, including *FDA Consumer.*

☑ Food and Nutrition Information Center (FNIC)
<http://www.nal.usda.gov/fnic/>

FNIC is part of the USDA and the Agricultural Research Service (ARS) in partnership with the University of Maryland and Howard University. Site features include FNIC databases, topics A to Z, dietary guidelines, and more. Information specific to food safety and security includes training materials to prevent foodborne illnesses, Foodsafe discussion group, and links to various food safety resources.

Food Navigator
<http://www.foodnavigator.com>

This Europe-based resource provides breaking news about foods, beverages, and a "large range of ingredients and additives including: nutraceuticals, vitamins and minerals, flavors, enzymes, colors, emulsifiers, fibers, preservatives and hydrocolloids." Site features include industry news, legislation, market reports, daily and weekly newsletters, and events.

Global Health.gov
<http://www.globalhealth.gov/>

This site was developed by the Office of Global Health Affairs, part of the U.S. Department of Health and Human Services. Site features include calendar of events, global health headlines, statements and speeches, reports and publications, country information, world health statistics, fact sheets, surgeon general's page, and links to useful sites. Information about food safety and security includes fact sheets about the food supply, travel advisories, and various global health topics.

Lonely Planet Health
<http://www.lonelyplanet.com/health/>

Lonely Planet is a site dedicated to exploration and travel. The health section of the Lonely Planet site includes information about staying healthy when traveling by being careful about food and drink, water purification, and nutrition.

☑ MayoClinic
<http://www.mayoclinic.com/>

The renowned Mayo Clinic has locations in Rochester, Minnesota; Jacksonville, Florida; and Scottsdale, Arizona. The Mayo Clinic Web site offers authoritative consumer health information, including diseases and conditions, healthy living, drug search, health tools such as calculators and quizzes, books and newsletters, and more.

☑ MedlinePlus
<http://www.medlineplus.gov/>

MedlinePlus, produced by the National Library of Medicine (NLM), provides access to authoritative, full-text information on more than 500 diseases and conditions of interest to consumers and health professionals alike. Resources include a medical encyclopedia and dictionaries, downloadable health pamphlets, health information from the media, information on prescription and nonprescription drugs, and links to external sites with reliable health content. The content is presented in the English and Spanish languages.

☑ National Agricultural Library (NAL)
<http://www.nal.usda.gov/>

NAL is one of the national libraries run by the U.S. government. Its mission is to "increase the availability and utilization of agricultural information for researchers, educators, policymakers, consumers of agricultural products, and the public." Site features include AgNIC, the AGRICOLA database, various publications pertinent to food safety and security, services, programs, and a calendar of events.

Nutrition Information and Resource Center (NIRC)
<http://nirc.cas.psu.edu/index.cfm>

The College of Agricultural Sciences at Pennsylvania State University uses this site to provide information about general nutrition, food science, nutrition news, and community resources. Information specific to food safety and security includes materials in the NIRC Library, online publications, and links to external resources, such as the USDA, NAL, and CDC.

SciTech Resources
<http://www.scitechresources.gov/>

The U.S. Department of Commerce lists government Web sites related to science and technology including food, agriculture, biology, health, medicine, and other areas. Sites are searchable by keyword, topic, agency, or resource type.

Seafood Information Network (SeafoodNIC)
<http://seafood.ucdavis.edu/home.htm>

SeafoodNIC is a site maintained by the University of California at Davis. Site features include guidelines and regulations, links and information sources, nutrition and food labeling, publications, sanitation, online news and electronic mailing lists, hazard analysis information, events, and other information.

State Extension Programs
<http://www.co.bay.mi.us/bay/home.nsf/Public/State_Extension_
   Programs.htm>

Bay County, Michigan, maintains this list of land grant universities and state extension programs. State extension offices focus on environmental and agricultural issues, including topics related to food safety and security.

UC-Davis Vegetable Research and Information Center
<http://vric.ucdavis.edu/>

The purpose of the Vegetable Research and Information Center (VRIC) at the University of California at Davis is to "foster appropriate research, collect and disseminate information relevant to consumers, growers and processors in the California vegetable industry." Site features include a virtual tour of vegetable production areas in the state of California; issues, news, and events; vegetable information by crop, topic, and postharvest; and links to external sites. Some site information is featured in the Spanish language.

Chapter 3

Major Safety and Security Sites

The sites included in this section cover major safety and security issues. Consult the more specific chapters in this book to focus on individual diseases, conditions, issues, concerns, or controversies related to food safety and security. The symbol ☑ is used to denote major resources with authoritative and original content.

☑ American Red Cross
<http://www.redcross.org/>
<http://www.prepare.org/>

The American Red Cross (ARC) devotes itself to providing "relief to victims of disasters and [helping] people prevent, prepare for, and respond to emergencies." Site features include news, services, and publications. Information specific to food safety and security describes food safety during power outages, terrorism preparedness, and food assistance. ARC's prepare.org site focuses on disaster preparedness information for the disabled, elderly, pets and pet owners, children, and so forth.

Bureau of Industry and Security
<http://www.bxa.doc.gov/>

The Bureau of Industry and Security (BIS) is part of the U.S. Department of Commerce. The mission of BIS is to advance U.S. national security, foreign policy, and economic interests. Site features include policies and regulations, licensing, compliance and enforcement, seminars and training, and international programs.

Commission on National Security
<http://www.nssg.gov/>

This commission is responsible for reviewing the national security policies and processes of the United States and forecasting alternative scenarios through 2025. Site features include its charter, commission members, and reports.

Consumer Product Safety Commission (CPSC)
<http://www.cpsc.gov/>

This U.S.-based agency focuses on the safety of durable consumer goods such as appliances, household wares, and the like that may have implications for food safety. The site includes recalls and product safety news, poison prevention, publications, e-mail announcements, and other information. Some of the site content is available in the Spanish language.

Environmental Protection Agency (EPA)
<http://www.epa.gov/>

The U.S. Environmental Protection Agency (EPA) is "responsible for a number of activities that contribute to food security within the United States, in areas such as food safety, water quality, and pesticide applicator training." The EPA protects human health and safeguards the environment by monitoring air, water, and land. The site includes useful information for laypersons and experts related to legislation, regulations, publications, software, and more.

Federal Emergency Management Agency (FEMA)
<http://www.fema.gov/>

FEMA is a U.S. agency that focuses on disaster preparedness and emergency management issues including news reports, training programs, weather warnings, and disaster assistance information. Current weather, storm watch, disaster updates, disaster fact sheets, agency news, and flood maps are a few of the many features of this site.

National Institute for Occupational Safety and Health (NIOSH)
<http://www.cdc.gov/niosh/homepage.html>

NIOSH focuses on the prevention of work-related illnesses and promotion of occupational safety. Site features include eNews, publications, databases, hazard evaluations, training, conferences, and interactive tools such as the National Agriculture Safety Database (NASD).

National Security Agency (NSA)
<http://www.nsa.gov/>

The NSA is a U.S. government agency that collects and analyzes sensitive intelligence information. The public information portion of the NSA site includes press releases, briefings, reports, declassified information, and congressional testimony.

National Security Research Area
<http://www.rand.org/natsec_area/>

The RAND Corporation is a nonprofit organization that provides "objective analysis and effective solutions that address the challenges facing the public and private sectors around the world." RAND's National Security Research and Analysis division focuses on some areas related to food safety including food-chain threats, agroterrorism, emerging infectious diseases, and terrorism preparedness. Site features include policies, practices, publications, and links to external sites.

Occupational Safety and Health Administration (OSHA)
<http://www.osha.gov/>

OSHA is part of the U.S. Department of Labor and was established in 1971 to "prevent work-related injuries, illnesses, and deaths." Site features include inspection data, statistics, directives, regulations, and a mechanism for workers to file complaints electronically.

Ready.gov
<http://ready.gov/>

The U.S. Department of Homeland Security, which was formed about a month after September 11, 2001, provides useful information on its Ready.gov site. Site features include information about biological, chemical, nuclear, and radiation threats as well as tips on making an emergency kit (food, water, first aid supplies, and the like) and developing an emergency plan.

Chapter 4

Food Safety and Security Sites

The sites included in this section cover many aspects of food safety and security and are excellent resources for exploring various topics related to staying healthy while preparing, eating, and storing foods. Some sites that focus on "food security" define this term as fighting the poverty that causes hunger, rather than referring to safe food supplies. Consult the more specific chapters in this book to focus on individual issues, concerns, or controversies related to the complexities of food safety and security. The symbol ☑ is used to denote major resources with authoritative and original content.

Arctic Health—Food Safety
<http://www.arctichealth.org/healthtopics3.php?topic_id=75>

This Arctic Health site is a product of the National Library of Medicine (NLM) and the University of Alaska at Anchorage. The site features evaluated and annotated links to "hundreds of local, state, national, and international agencies, as well as from professional societies and universities." Subjects related to food safety include fish poisoning, infant botulism, health advisories related to mercury, marine toxins, and other information.

Biotech Food Safety
<http://groups.ucanr.org/sbc/Outreach/Biotech_Food_Safety.htm>

The University of California Seed Biotechnology Center has organized this list of links related to food safety. Topic areas include transgenic crops, genetically modified foods, government organizations, nongovernment organizations, universities, feed industry, food industry, mycotoxins, and so on. Site features include news; resources; and links to industry, government, and university sites.

☑ Center for Food Safety and Applied Nutrition (CFSAN)
<http://vm.cfsan.fda.gov/>

CFSAN is one of six product centers charged with carrying out the mission of the U.S. Food and Drug Administration (FDA). CFSAN promotes and protects "the public's health by ensuring that the nation's food supply is safe, sanitary, wholesome, and honestly labeled, and that cosmetic products are safe and properly labeled." The CFSAN site includes information about selected topics such as acrylamide, bovine spongiform encephalopathy (mad cow disease), foodborne illness, food labeling, and more (see Figure 4.1).

Center for Food Safety Engineering (CFSE)
<http://www.cfse.purdue.edu/index.cfm>

Purdue University's CFSE focuses on multidisciplinary research related to food safety engineering. Site features include news, events, promotional media, annual reports, and external links to similar resources.

Department of Health and Human Services—Drug and Food Information
<http://www.hhs.gov/drugs/index.shtml>

The Department of Health and Human Services is the "United States government's principal agency for protecting the health of all Americans and providing essential human services, especially for those who are least able to help themselves." This site provides useful information about drug and food advisories, alerts, and recalls; safe drinking water; dietary supplements; and food labeling, preparation, irradiation, and allergies.

Food Safety and Sanitation
<http://www.fiu.edu/~nutreldr/OANP_Toolkit/Food_Safety02_18_03.htm>

Florida International University's National Policy and Resource Center on Nutrition and Aging provides this section on food safety which includes food handling standards, recalls, outbreaks, monitoring, and quality improvement. This information is part of a larger electronic work titled *Older American Nutrition Program Toolkit.*

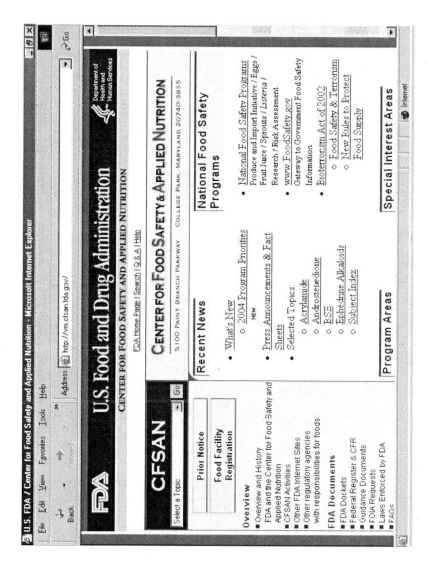

FIGURE 4.1. Center for Food Safety and Applied Nutrition (CFSAN) <http://vm.cfsan.fda.gov/>

Food Safety Consortium
<http://www.uark.edu/depts/fsc/>

This consortium was formed by the University of Arkansas, Iowa State University, and Kansas State University. Site features include the consortium newsletter, annual report, personnel directory, research projects, food safety presentations, current events, news, and more.

Food Safety Network
<http://www.foodsafetynetwork.ca/food.htm>

Canada's Food Safety Network uses "safe food, from farm to fork" as its motto. Site features include annotated links related to food additives, pathogens, and production; food codes and regulations; foodborne illnesses; food safety at home; food security; hazard assessment; handwashing; outbreak surveillance; food irradiation; restaurant inspection; seafood; vegetable production; and bottled water.

☑ Food Safety Project
<http://www.extension.iastate.edu/foodsafety/>

Iowa State University Extension focuses on "food safety from farm to table." This project site includes information about food irradiation, education, training, foodborne pathogens, and legislation. Site features include hot topics such as bird flu, acrylamide, and mad cow disease; food safety news; fact sheets; and more.

☑ Food Safety Research Information Office (FSRIO)
<http://www.nal.usda.gov/fsrio/>

The National Agricultural Library's Food Safety Research Information Office (FSRIO) provides information for consumers, researchers, and industry workers. Site features include news; research expenditures by agency and year; legislation; annual reports by agency; food safety briefing papers by agency; and links to food-related journals, organizations, topics, and universities involved with agricultural research (see Figure 4.2).

☑ Food Safety Risk Analysis Clearinghouse
<http://www.foodriskclearinghouse.umd.edu/>

The FDA and the University of Maryland formed the Joint Institute for Food Safety and Applied Nutrition (JIFSAN). Their Food Safety Risk Analysis Clearinghouse site is organized into information

FIGURE 4.2. Food Safety Research Information Office (FSRIO) <http://www.nal.usda.gov/fsrio/>

categories including risk analysis, community, nutrition and labeling, researcher tools, hazards, databases, commodity, resource type, consumer information, and risk assessment consortium. Special features include hot topics (acrylamide, dioxin, bovine spongiform encephalopathy) and events (see Figure 4.3).

Food Safety Throughout the Food System
\<http://foodsafety.cas.psu.edu/\>

Pennsylvania State University's food safety site focuses on prevention with its Food Safety Database, news updates, rumor control, information about recalls, discussion groups, links to various publications, tips about home food preservation, educational materials, and events.

Food Security in the United States
\<http://www.ers.usda.gov/briefing/foodsecurity/\>

The Economic Research Service (ERS) of the U.S. Department of Agriculture (USDA) provides this overview of food security in the United States. Points include how to measure household food security; community food security, conditions and trends, and more. In this case, the term food security refers to food insecurity or hunger caused by poverty.

Food Security Resources
\<http://www.fns.usda.gov/fsec/\>

The USDA's Food and Nutrition Service (FNS) provides access to food security action resources and measurement including statistical reports, handbooks, briefings, and grants. Publications include *The National Nutrition Safety Net: Tools for Community Food Security,* and *A Guide to Measuring Household Food Security*. In this case, the term food security refers to food insecurity or hunger caused by poverty.

☑ Food Standards Australia New Zealand
\<http://www.foodstandards.gov.au/\>

Food Standards Australia New Zealand "protects the health and safety of the people in Australia and New Zealand by maintaining a safe food supply." Site features include standards, industry assistance, recalls, food surveillance, publications, and more. Information specifically geared toward consumers includes acrylamide, antibiotics in food, avian influenza, food additives, food irradiation, food labeling,

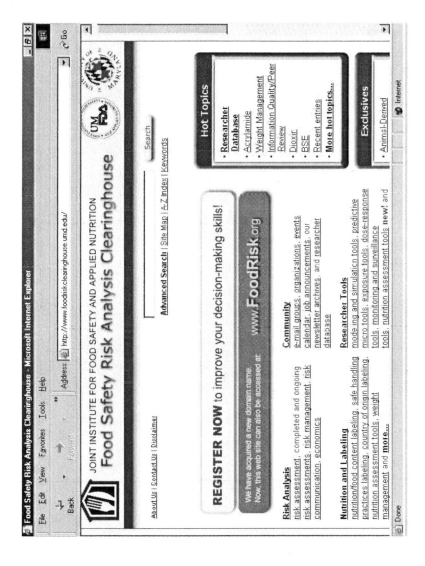

FIGURE 4.3. Food Safety Risk Analysis Clearinghouse <http://www.foodriskclearinghouse.umd.edu/>

genetically modified foods, listeria, mad cow disease, mercury in fish, pregnancy and food, and more.

☑ Foodlink
<http://www.foodlink.org.uk/>

The United Kingdom's Food and Drink Federation organizes and maintains the Foodlink site. Site features include an alphabetical list of food safety topics, factfiles, quizzes, and resources. Highlights include an image bank, downloadable files to be used to promote food safety events, school poster competitions, and more (see Figure 4.4).

FoodSafe Program
<http://foodsafe.ucdavis.edu/>

The University of California at Davis maintains a FoodSafe program that focuses on consumer and industry information related to food safety. Site features include hot topics, food safety experts, consumer advice, food industry sources, hazard analysis information, food safety music, and more.

☑ FoodSafety.gov
<http://www.foodsafety.gov/>

This gateway to U.S. government information about food safety includes news; safety alerts; content for children, teenagers, and educators; reporting of illnesses and complaints; foodborne pathogens; state and federal agencies; and food industry assistance (see Figure 4.5).

Home Food Safety
<http://www.homefoodsafety.org/index.jsp>

This site is a joint project between the American Dietetic Association (ADA) and the ConAgra Foods Foundation (see Figure 4.6). Site features include tips (wash hands; separate raw and ready-to-eat foods; cook foods to proper temperatures; and refrigerate promptly); news; an Interactive Kitchen, which uses nine questions to test food-handling knowledge; content in the Spanish language; and links to external sites such as the USDA, CFSAN, and Centers for Disease Control and Prevention (CDC).

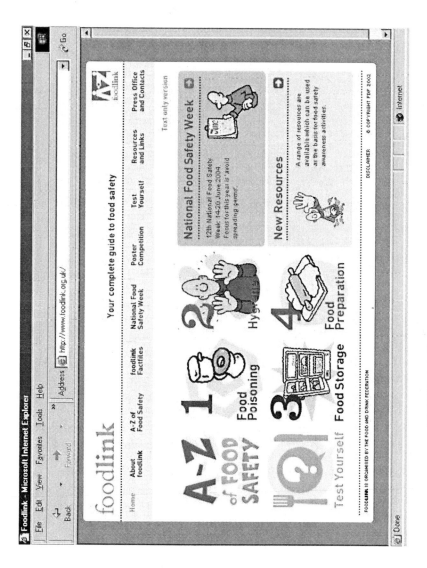

FIGURE 4.4. Foodlink <http://www.foodlink.org.uk/>

29

FIGURE 4.5. FoodSafety.gov <http://www.foodsafety.gov/>

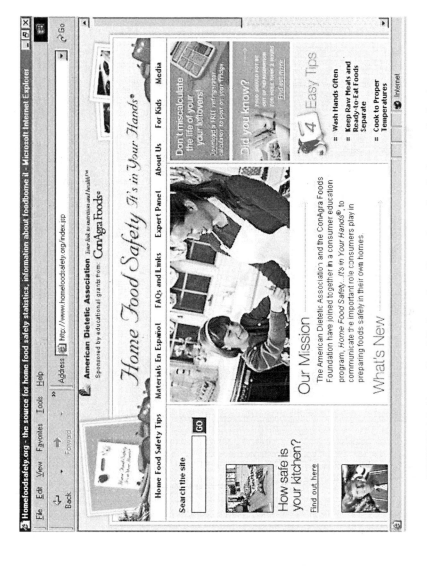

FIGURE 4.6. Home Food Safety <http://www.homefoodsafety.org/index.jsp>

Household Food Security—Community Nutrition
<http://www.fao.org/es/ESN/nutrition/household_emergencies_en.stm>

The Food and Agriculture Organization (FAO) of the United Nations works to prevent worldwide hunger. Site information about household food security includes using food-based approaches to agriculture, emergencies, home gardens, and more. Content is presented in English, French, and Spanish languages.

Iowa State Food Safety Project
<http://www.extension.iastate.edu/foodsafety/>

Iowa State University Extension's Food Safety project focuses on "research-based, unbiased information on food safety and quality" for consumers, educators and students. Site features include content and links related to biosecurity, irradiation, education, training, foodborne pathogens, news, hot topics, and more.

☑ MedlinePlus: Food Safety
<http://www.nlm.nih.gov/medlineplus/foodsafety.html>

MedlinePlus organizes content and links including latest news about food safety as well as general/overviews about handling food safely, research, specific conditions/aspects, dictionaries/glossaries, law and policy, and more links related to the subject (see Figure 4.7).

Programme of Food Safety and Food Aid
<http://www.who.int/foodsafety/en/>

The World Health Organization's content about food safety includes information about microbiological risks, chemical risks, genetically modified foods, food standards, foodborne diseases, production-to-consumption information, highlights, news, and publications.

Public Citizen Food Agriculture
<http://www.citizen.org/cmep/foodsafety/>

Founded in 1971, Public Citizen is a nonprofit organization that focuses on consumer advocacy, including food safety. Site features include information on water, food irradiation, safe school lunches, and more.

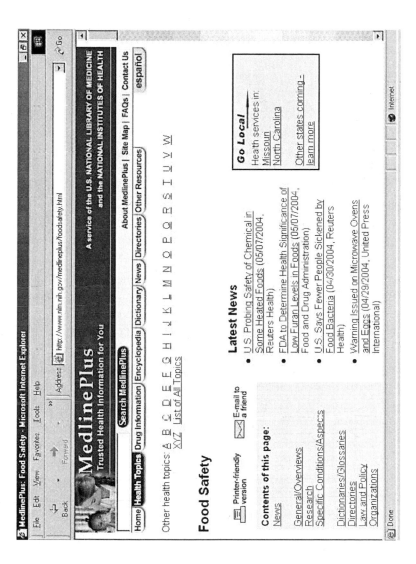

FIGURE 4.7. MedlinePlus: Food Safety <http://www.nlm.nih.gov/medlineplus/foodsafety.html>

USDA/FDA Foodborne Illness Education Information Center
<http://www.nal.usda.gov/fnic/foodborne/index.html>

The USDA and FDA combine their efforts in this foodborne illness education information center. Their materials can be used to educate the general public about food safety. Site features include a discussion group; database of educational materials; external links; and food safety education stories, solutions, and strategies, which share compelling approaches from other institutions. Materials can be searched by audience, language, and format.

Chapter 5

Diseases and Conditions

The sites included in this section provide information about specific diseases and conditions caused by unsafe foods and situations. The symbol ☑ is used to denote major resources with authoritative and original content.

ANAPHYLAXIS

Anaphylaxis Campaign
<http://www.anaphylaxis.org.uk/>

The United Kingdom's Anaphylaxis Campaign provides up-to-date information for persons with life-threatening allergies, including food alerts, and airlines that have agreed to remove peanut products from their meals.

Food Allergies
<http://edition.cnn.com/HEALTH/library/HQ/00709.html>

This information is hosted on the CNN.com site but was developed by MayoClinic.com. Tips include symptoms of allergic reactions, allergy tests, dangers of air travel for persons with peanut allergies, and packing for emergencies.

Food Allergy and Anaphylaxis Network (FAAN)
<http://www.foodallergy.org/>

FAAN provides useful information about recalls, alerts, common food allergens, a calendar of events, and more.

Travel Safety Tips—Severe Allergic Reactions
<http://www.hotelfun4kids.com/travelsafetytips/severeallergy.
  htm>

The Hotelfun4kids site is devoted to family travel. The information about food allergies includes information about symptoms, causes, and prevention of anaphylactic shock; traveling with severe allergies; and links to external sites.

CHRONIC ILLNESS

Food Safety—GMHC's Nutrition and Wellness Program
<http://www.gmhc.org/health/nutrition/factsheets/food_safety.
  html>

The Gay Men's Health Crisis (GMHC) developed this understandable list of food-handling tips, including handwashing, avoiding cross-contamination, thawing foods safely, and other handling practices that are applicable to immunocompromised individuals.

Food Safety for Persons with AIDS
<http://www.fsis.usda.gov/Fact_Sheets/Food_Safety_for_Persons_
  with_AIDS/index.asp>

The U.S. Department of Agriculture's (USDA) Food Safety Inspection Service (FSIS) provides understandable information about dangerous organisms (salmonella, listeria, campylobacter, etc.), foods to avoid, food handling, food storage, and other tips.

Food Safety for the Chronically Ill
<http://hgic.clemson.edu/factsheets/HGIC3643.htm>

Clemson University's Home and Garden Information Center offers useful information about the dangers of specific foods for persons who are chronically ill. Advice relates to shopping, kitchen preparation, cooking, eating in restaurants, and traveling abroad. The information is derived from the USDA's Food Safety Information Service.

Food Safety Guidelines for Immunosuppressed Patients
<http://www.library.umc.edu/pe-db/pe-food_safety_immuno.html>

The University of Mississippi Medical Center's Department of Nutrition offers this practical advice for immunosuppressed patients including safe food handling, microwave cooking, and grocery shopping. The content was adapted from information developed by the Seattle Cancer Care Alliance and Fred Hutchinson Cancer Research Center.

DISEASE OUTBREAKS AND SURVEILLANCE

☑ CDC Green Sheet
<http://www2.cdc.gov/nceh/vsp/VSP_RptGreenSheet.asp>

The Centers for Disease Control and Prevention (CDC) is a U.S. federal agency located in Atlanta, Georgia, that focuses on protecting health and safety by monitoring and preventing disease. The CDC publishes *Summary of Sanitation Inspections of International Cruise Ships,* also known as the "Green Sheet." This publication lists ships by name, inspection date, and score. Sanitation standards cover water; food preparation and holding; potential contamination of food; and general cleanliness, storage, and repair.

Chemical and Physical Contaminants
<http://peaches.nal.usda.gov/foodborne/fbindex/Contaminants.
asp>

The USDA/Food and Drug Adminstration (FDA) Foodborne Illness Education Center provides information about chemical and physical contaminants including links to government sites and nongovernment sites related to contaminants such as acrylamide, aflatoxins, foreign objects, dioxins, harmful algae, residues, mycotoxins, and other substances.

European Disease Outbreaks
<http://www.defra.gov.uk/animalh/diseases/european_news.htm>

The United Kingdom's Department of Environment, Food, and Rural Affairs (DEFRA) concerns itself with economic, social, and environmental aspects of sustainable development. Site features include documentation of disease outbreaks in Europe; notifiable diseases; diseases spread from animals to humans; diseases, monitoring, surveillance, and control; and other information.

☑ Food Contamination and Poisoning
<http://www.nlm.nih.gov/medlineplus/
   foodcontaminationpoisoning.html>

MedlinePlus organizes content and links including the latest news about foodborne illnesses caused by contamination or poisoning, as well as general overviews, clinical trials, diagnosis/symptoms, pictures/diagrams, prevention/screening, specific conditions/aspects, organizations, content in other languages, and more links related to the subject.

☑ Food Safety While Hiking, Camping, and Boating
<http://www.fsis.usda.gov/OA/pubs/hcb.htm>

The FSIS "is the public health agency in the U.S. Department of Agriculture responsible for ensuring that the nation's commercial supply of meat, poultry, and egg products is safe, wholesome, and correctly labeled and packaged." This document explains methods of keeping food and water safe to consume while hiking, camping, and boating.

Foodborne Disease
<http://www.osha.gov/SLTC/foodbornedisease/index.html>

The U.S. Department of Labor's Occupational Safety and Health Administration (OSHA) offers annotated links to external foodborne disease sites and information about botulism. Categories include OSHA standards; hazard recognition, evaluation, and investigation; control and prevention; hot topics; and news releases.

☑ Foodborne Diseases
<http://www.niaid.nih.gov/factsheets/foodbornedis.htm>

The National Institute of Allergy and Infectious Diseases of the U.S. National Institutes of Health (NIH) developed this fact sheet related to typical foodborne illnesses including botulism, campylobacteriosis, *E. coli* infections, salmonellosis, shigellosis, and so on.

Foodborne Illness
<http://www.cdc.gov/ncidod/dbmd/diseaseinfo/ foodborneinfections_g.htm>

The CDC is a U.S. federal agency located in Atlanta, Georgia, that focuses on protecting health and safety by monitoring and preventing disease. This information about foodborne illness is provided in English and Spanish languages. The content covers common infections (campylobacter, salmonella, *E. coli*), diagnosis, treatment, consulting a doctor, number of yearly cases, disease surveillance and outbreaks, and methods to prevent foodborne illness.

☑ Foodborne Illness Education Information Center
<http://www.nal.usda.gov/fnic/foodborne/>

The USDA and FDA combine their efforts in this information center to provide materials that can be used to educate the general public about food safety. Site features include a discussion group; database of educational materials; external links; and food safety education stories, solutions, and strategies which share compelling approaches from other institutions. Materials can be searched by audience, language, and format (see Figure 5.1).

☑ FoodNet
<http://www.cdc.gov/foodnet/default.htm>

FoodNet is the CDC's surveillance network for foodborne diseases. Site features include an interactive map of health departments, diseases and pathogens under surveillance, reports and studies, data sources, and what's new.

The entry for foodborne infections includes general information (definitions, diagnosis, treatment); technical information (etiology, incidence, transmission, trends); and additional information (links to other information developed by the CDC). Some site content is presented in the Spanish language.

An estimated 76 million cases of foodborne disease occur each year in the United States.

Source: DBMD—Foodborne Infections—General Information <http://www/cdc. gov/ncidod/dbmd/diseaseinfo/foodborneinfections_g.htm>.

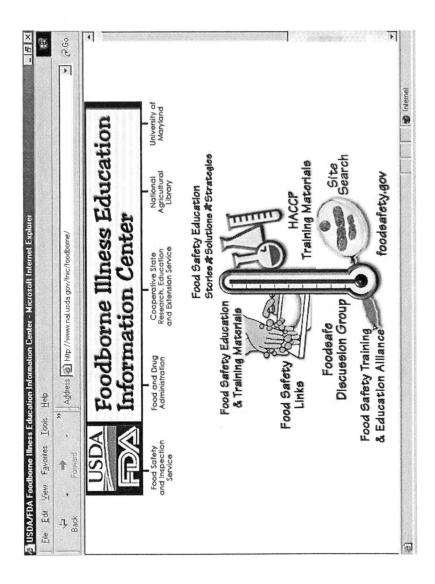

FIGURE 5.1. Foodborne Illness Education Information Center <http://www.nal.usda.gov/fnic/foodborne/>

40

Organisms That Can Bug You
<http://www.fightbac.org/bug.cfm>

The Partnership for Food Safety Education is a U.S.-based public-private partnership that strives to educate the public about safe food-handling practices. The site includes a chart of organisms, sources of illness, and symptoms.

☑ Outbreak Response and Surveillance Unit
<http://www.cdc.gov/foodborneoutbreaks/>

The U.S. CDC monitors outbreaks of foodborne diseases. Site features include an investigation toolkit (questionnaire, foodborne disease finder, guidelines, training materials); outbreak reports and publications; disease report forms and reporting requirements; atlas of food consumption habits; and patient information about bacterial, chemical, parasitic, and viral causes of foodborne illnesses.

Parasites and Foodborne Illness
<http://www.fsis.usda.gov/oa/pubs/parasite.htm>

This information defines terms; explains the role of parasites in illnesses caused by foods; and describes symptoms, risks, and prevention of diseases such as trichinosis, giardiasis, toxoplasmosis, cyclosporiasis, and tapeworms.

☑ Risks from Food and Drink
<http://www.cdc.gov/travel/food-drink-risks.htm>

The Travelers' Health site maintained by the National Center for Infectious Diseases of the CDC includes information about risks from food and drink and methods of treating water to make it safe to drink.

MAD COW DISEASE

Bovine Spongiform Encephalopathy (BSE)
<http://www.who.int/emc/diseases/bse/>

The World Health Organization (WHO) is a specialized health agency within the United Nations. Its Communicable Disease Surveillance and Response (CSR) division developed this information about bovine spongiform encephalopathy (mad cow disease), including a fact sheet, surveillance and control, and information resources.

☑ Bovine Spongiform Encephalopathy
<http://www.fda.gov/oc/opacom/hottopics/bse.html>

The FDA organized this information related to bovine spongiform encephalopathy (mad cow disease). Site information includes general background, recent actions, consumer information, industry/veterinary information, bovine-based vaccines, and blood safety.

BSE and CJD Information and Resources
<http://www.cdc.gov/ncidod/diseases/cjd/cjd.htm>

The CDC developed this information about bovine spongiform encephalopathy, a disease in cattle that is similar to Creutzfeldt-Jakob Disease, a neurodegenerative disorder in humans.

☑ CDC's Traveler's Health Information on Bovine Spongiform Encephalopathy
<http://www.cdc.gov/travel/diseases/madcow.htm>

The CDC developed this information on bovine spongiform encephalopathy, including a disease description, occurrence, risk for travelers, and prevention.

Mad Cow Disease/Mad Deer Disease
<http://www.organicconsumers.org/madcow.htm>

Organic Consumers Association (OCA) is a nonprofit organization concerned with "food safety, industrial agriculture, genetic engineering, corporate accountability, and environmental sustainability." Their site information about mad cow disease and mad deer disease includes articles, daily news, and links to external sites.

Priondata.com
<http://www.priondata.org/>

This site is sponsored by Microsens Biotechnologies, which manufactures a reagent for prions. Site features include headlines, news, background information, and links related to diseases caused by prions, such as a neurodegenerative disease.

PREGNANCY

Agency Updates Advice to Pregnant and Breastfeeding Women on Eating Certain Fish
<http://www.food.gov.uk/news/pressreleases/2003/feb/tuna_mercury>

The United Kingdom's Food Standards Agency offers this advice to pregnant and breast-feeding women about the dangers of eating too much tuna, shark, swordfish, or marlin. This document explains the suggested limits for fish consumption in this at-risk group and details mean levels of methylmercury.

During Your Pregnancy: Food Safety
<http://www.marchofdimes.com/pnhec/159_826.asp>

The March of Dimes is a nonprofit organization dedicated to preventing birth defects and reducing infant mortality through research and education. Their site information about food safety during pregnancy covers foods to avoid.

Food Safety in Pregnancy
<http://www.nzfsa.govt.nz/consumers/food-safety-topics/foodborne-illnesses/pregnancy/index.htm>

The New Zealand Food Safety Authority (NZFSA) developed this sixteen-page color brochure that "outlines the different types of foodborne illness, how they can affect pregnant women and their babies and how women can help to avoid them."

Pregnancy and Food Safety
<http://healthlink.mcw.edu/article/949098908.html>

The Medical College of Wisconsin's HealthLink site provides concise information about food risks during pregnancy. The information was developed by a registered dietitian.

Protect Your Unborn Baby
<http://www.cspinet.org/foodsafety/brochure_pregnancy.html>

Center for Science in the Public Interest (CSPI) is a Washington, DC-based nonprofit organization devoted to consumer advocacy in several areas of public health. Their site offers tips for avoiding food poisoning, especially during pregnancy.

What You Need to Know About Mercury in Fish and Shellfish
<http://www.epa.gov/waterscience/fishadvice/advice.html>

The U.S. Environmental Protection Agency (EPA) and FDA developed this information about dangers of consuming fish and shellfish while pregnant, nursing, or in early childhood. The recommendations focus on limiting consumption of specific types of fish (shark, tilefish, swordfish, and king mackerel) because of elevated mercury levels; consuming fish with lower levels of mercury; and abiding by local fish advisories. This content is also featured in the Spanish language.

Chapter 6

Specific Issues, Concerns, and Controversies

The sites included in this chapter cover specific issues, concerns, and controversies related to food safety and security. See also the section about specific diseases and conditions. The symbol ☑ is used to denote major resources with authoritative and original content.

ADDITIVES, DYES, EMULSIFIERS, FORTIFIERS, FLAVORS, COLORS, AND PRESERVATIVES

Additives in Meat and Poultry Products
<http://www.fsis.usda.gov/OA/pubs/additive.htm>

The Food Safety Inspection Service (FSIS) "is the public health agency in the U.S. Department of Agriculture responsible for ensuring that the nation's commercial supply of meat, poultry, and egg products is safe, wholesome, and correctly labeled and packaged." This document includes information about history, definition, monitoring, regulation, and labeling requirements of food additives.

Compendium of Food Additive Specifications
<http://apps3.fao.org/jecfa/additive_specs/foodad-q.jsp>

The Joint FAO/WHO Expert Committee on Food Additives (JECFA) developed this database of food additives (other than flavors). FAO is the Food and Agriculture Organization of the United Nations, and WHO is the World Health Organization. The content is available in Arabic, French, Spanish, and English languages. Entries can be searched by substance name, International Numbering System (INS) number, Chemical Abstracts Service (CAS) number, functional use group (antioxidants, emulsifiers, stabilizers, etc.), purity group (cadmium, lead, arsenic, etc.), and food additives designated as tentative. Typical database entries include substance name, synonyms, chemical name(s), chemical formula, structural formula, description, characteristics, method of assay, and more.

EU Proposals on Fortified Foods
<http://www.epha.org/a/9>

The European Public Health Alliance (EPHA) site explains the European Union's (EU) draft directives on fortified foods. The site also includes information about misleading food claims, labeling, and more.

Food Additives and Flavourings
<http://europa.eu.int/comm/food/fs/sfp/flav_index_en.html>

"The European Commission embodies and upholds the general interest of the [European] Union and is the driving force in the Union's institutional system." The site offers definitions, explanations, legislation, and directives related to food additives and flavors.

Food Fortification Safety and Legislation
<http://www.unu.edu/unupress/food/V192e/ch04.htm>

United Nations University (UNU) Press publishes a number of scholarly publications, including *Food and Nutrition Bulletin.* The information about safety and legislation of fortified foods includes trends in food hygiene regulation, sampling and analysis at import, technology and quality control, general principles for addition of nutrients to foods, and quality assurance and control.

Food Standards Agency—Food Additives
<http://www.foodstandards.gov.uk/safereating/additivesbranch/>

The United Kingdom's Food Standards Agency was established by Parliament in 2000 "to protect the public's health and consumer interests in relation to food." The site section about additives explains food preservatives, colors, emulsifiers, flavors, sweeteners, and similar substances, in addition to discussing associated legislation and practices.

Ingredients, Food Additives, and Nutrition
<http://www.foodnavigator.com/>

This Europe-based resource provides breaking news about foods and beverages, covering a "large range of ingredients and additives including: nutraceuticals, vitamins and minerals, flavors, enzymes, colors, emulsifiers, fibers, preservatives and hydrocolloids."

Joint Expert Committee on Food Additives (JECFA)—
Monographs and Evaluations
<http://www.inchem.org/pages/jecfa.html>

The Canadian Centre for Occupational Health and Safety (CCOHS) and the International Programme on Chemical Safety cooperate to provide access to "peer-reviewed chemical safety-related publications and database records from international bodies." These publications are part of the FAO Nutrition Report Series, and cover evaluating the toxicology and purity of various substances, including antimicrobials, antioxidants, and other food additives.

MedlinePlus: Food Additives
<http://www.nlm.nih.gov/medlineplus/ency/article/002435.htm>

This medical encyclopedia entry on the MedlinePlus site includes definitions, functions, sources, side effects, and recommendations for food additives.

ADVISORIES, ALERTS, RECALLS, AND WARNINGS

FDA Recalls, Market Withdrawals and Safety Alerts
<http://www.fda.gov/opacom/7alerts.html>

The Food and Drug Administration (FDA) is part of the Department of Health and Human Services of the U.S. government and is responsible for "protecting consumers and promoting public health." The FDA site features recalls, withdrawals, and alerts that have been issued in the past sixty days.

Index of Canadian Food Recalls
<http://www.inspection.gc.ca/english/corpaffr/recarapp/recaltoce. shtml>

The Canadian Food Inspection Agency "delivers all federal inspection services related to food; animal health; and plant protection." The site provides e-mail notification of recalls and allergy alerts. Other features include current alerts and warnings, an index of food recalls, food recall archives, and warnings and advisories from Health Canada. Information displayed can be sorted by commodity (dairy, eggs, feeds, fertilizers, fish and seafood, fresh fruits and vegetables, grains, honey, potatoes) or key topic (animal health, biotechnology, horticulture, food recalls, processed foods).

☑ Recall Information Center FSIS Recalls
<http://www.fsis.usda.gov/OA/Fsis_recalls/index.asp>

This government information about food recalls includes definitions; frequently asked questions; procedures; archives of recall cases; state agencies; and state cases of contaminated, adulterated, or misbranded foods (see Figure 6.1).

RecallWarnings
<http://www.recall-warnings.com/>

The RecallWarnings site categorizes recall information and warnings related to a wide range of consumer products. The food and drugs section covers medications, vitamins, supplements, and recalls enacted by the FDA and the FSIS.

CARCINOGENS

The Cancer Project
<http://www.cancerproject.org/>

The Cancer Project is a program launched by the Physicians Committee for Responsible Medicine, "a nonprofit organization that promotes preventive medicine, conducts clinical research, and encourages higher standards for ethics and effectiveness in research." The

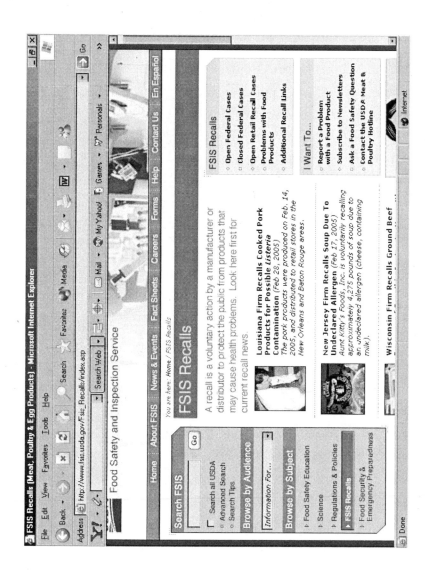

FIGURE 6.1. Recall Information Center FSIS Recalls <http://www.fsis.usda.gov/Fsis_Recalls/index.asp>

aim of the project is to prevent cancer and to improve survival after cancer has been diagnosed. The site includes information about food as medicine, research initiatives, ask the dietitian, cancer commentaries, and more. The food as medicine section explains specific nutrients (lycopene, carotenoids, vitamins, selenium, etc.) that promote health and may prevent cancer.

Cancer-Causing Factors
<http://www.hoag.org/CancerCenter/CancerCausingFactors.html>

Hoag Cancer Center in Newport Beach, California, explains cancer-causing factors, including ultraviolet radiation, ionizing radiation, industrial chemicals, diet, and tobacco.

Food Constituents: Natural Carcinogens
<http://interactive.usask.ca/ski/agriculture/food/foodnut/constit/constit_add5.html>

Saskatchewan Interactive is a Canadian site maintained by the Saskatchewan Centre for Soils Research at the University of Saskatchewan in Saskatoon. Site information about natural carcinogens includes definitions of terms; harmful substances produced by storage, cooking, and other methods; and colorful illustrations.

Natural Carcinogens and Anticarcinogens in America's Foods
<http://www.acsh.org/publications/booklets/nature.html>

The American Council on Science and Health (ACSH) is a nonprofit organization that concerns itself "with issues related to food, nutrition, chemicals, pharmaceuticals, lifestyle, the environment and health." The site includes a forty-one-page document about cancer-causing and cancer-fighting substances in foods. Written by physicians associated with reputable institutions, the publication defines and explains carcinogens and mutagens, aflatoxins and other mold toxins, toxins in common foods, cancer-fighting substances in foods, and hazards from natural and synthetic substances.

COOKED FOODS AND COOKING METHODS

Please also refer to the listings under the Kitchen Safety heading in this chapter.

Barbeque Food Safety
<http://www.fsis.usda.gov/factsheets/barbecue_food_safety/index.asp>

FSIS developed this information about safely purchasing, preparing, cooking, and storing foods that are cooked outdoors.

Cooking Safely in the Microwave Oven
<http://www.fsis.usda.gov/oa/pubs/fact_microwave.htm>

The USDA's FSIS developed this understandable information about cooking, defrosting, or reheating foods in a microwave oven.

Food Safety for Home Cooking
<http://www.lancaster.unl.edu/food/foodsafety.htm>

The University of Nebraska Cooperative Extension in Lancaster County organized this information about cooking at home. Some resources were developed by Lancaster County dietitians while others link to externally available information. Categories include advance preparation of food; refrigerator, freezer, and pantry storage; cooking times and temperatures; microwave, slow cooker, and grilling; specific foods; special situations; food safety basics; home canning and freezing; and food safety links.

Methods of Cooking
<http://ag.ansc.purdue.edu/meat_quality/cooking_methods.html>

The Purdue University Animal Sciences site provides information about safe cooking methods for meat, including broiling, roasting, braising, pot roasting, and microwaving.

DISASTERS AND EMERGENCIES

Disaster Assistance with Food
<http://vm.cfsan.fda.gov/~fsg/fsgdisas.html>

The Center for Food Safety and Applied Nutrition (CFSAN) is a department of the FDA, which in turn is part of the U.S. Department of Health and Human Services (HHS). Site information includes links to state and other federal government information about disaster preparation and postdisaster eating, cooking, and drinking.

Food Safety in a Power Outage
<http://www.prepare.org/basic/foodchart.htm>

Prepare.org is a disaster-preparedness site developed by the American Red Cross. The information about keeping food safely during and after a power outage is featured in normal and large fonts, and in the English, Arabic, Spanish, Farsi, French, Hmong, Japanese, Khmer, Korean, Lao, Russian, Tagalog, and Vietnamese languages. A chart explains whether specific foods can be kept, thawed, or discarded.

Is Food Safe After a Natural Disaster?
<http://www.ces.ncsu.edu/depts/foodsci/agentinfo/hot/natdis.html>

The North Carolina Cooperative Extension Service features annotated links to external information about disaster preparedness and its aftermath. This information can help consumers determine whether food and water is safe after a variety of disasters including power outages, floods, fire, hurricanes, tornados, and earthquakes.

Keeping Food Safe During an Emergency
<http://www.fsis.usda.gov/factsheets/keeping_food_Safe_during_an_emergency/index.asp>

This understandable fact sheet explains how to keep food safe during and after an emergency. Authoritative advice is supplemented with answers to a series of frequently asked questions and a chart listing specific foods to be kept or discarded.

FARMED FISH

Concerns Raised About Chemicals in Farmed Fish
<http://www.epha.org/a/991>

The European Public Health Alliance (EPHA) explains some recent research about carcinogens in salmon raised in European fish farms.

Farmed and Dangerous
<http://www.farmedanddangerous.org/>

The Coastal Alliance for Aquaculture Reform (CAAR) in Vancouver, Canada, developed this site to warn consumers about the dangers of consuming farmed fish. Site information includes the history of salmon farming, health effects of farmed fish, tools for activists interested in promoting wild salmon, reports, frequently asked questions, links, and more.

Intensive Marine Fish Aquaculture
<http://www.panda.org/downloads/marine/wwfaquaculturepolicy finaljuly2003.doc>

The World Wildlife Foundation (WWF) developed this position paper to explain the "environmental effects of intensive fish farming" including pollution, increased fish-feed production, genetically modified organisms, diseases, and other issues.

Mercury in Fish
<http://www.jifsr.gov/topics/tpmercury.htm>

The National Agricultural Library's (NAL) Food Safety Research Information Office (FSRIO) provides information for consumers, researchers, and industry workers. FSRIO's information about mercury in fish includes general information, fact sheets, reports, organizations, databases, publications, and events.

PCBs in Farmed Salmon
<http://www.ewg.org/reports/farmedPCBs/printversion.php>

The Environmental Working Group (EWG) is a "team of scientists, engineers, policy experts, lawyers and computer programmers" that examine "government data, legal documents, scientific studies" and other materials to discover potential environmental health threats and propose solutions. The information about farmed salmon discusses testing of farmed salmon purchased in Washington, DC; San Francisco; and Portland, Oregon, supermarkets.

FOOD IRRADIATION

Food Irradiation
<http://www.nal.usda.gov/fsrio/topics/tpfdirrad.htm>

The NAL's FSRIO provides information for consumers, researchers, and industry workers. FSRIO's information about food irradiation includes general information, fact sheets, reports, organizations, databases, publications, and events.

Food Irradiation—A Safe Measure
<http://www.fda.gov/opacom/catalog/irradbro.html>

The FDA explains the use of irradiation methods to prevent spoilage and reduce bacteria and pests in meats, fruits, and vegetables. This resource also answers common questions about the subject and details proper food-handling practices that can reduce or eliminate foodborne illnesses.

Food Irradiation—Frequently Asked Questions
<http://www.cdc.gov/ncidod/dbmd/diseaseinfo/foodirradiation. htm>

The U.S. Centers for Disease Control and Prevention (CDC) focuses on protecting health and safety by monitoring and preventing disease. Site information about food irradiation covers definitions; processes (gamma rays, electron beams, X-rays) used; effects on foods; measurement of irradiation doses; effects on microbes; approval and doses for specific foods; and other facts.

Food Irradiation Index
<http://www.hc-sc.gc.ca/food-aliment/fpi-ipa/e_irradiation_index. html>

Health Canada is Canada's federal health agency that works with provincial and territorial governments "to develop health policy, enforce health regulations, promote disease prevention and enhance healthy living for all Canadians." The site indexes information about industry requests to irradiate foods including shrimp, mangoes, ground beef, and poultry. Index information summarizes a specific request and its evaluation.

Irradiation
<http://www.foodsafety.gov/~fsg/irradiat.html>

FoodSafety.gov is the U.S. government's gateway to information about food safety. The information about food irradiation includes documents from the CDC, FDA, USDA, FSIS, and other agencies.

GENETIC ENGINEERING
AND GENETICALLY MODIFIED ORGANISMS

The Campaign to Label Genetically Engineered Foods
<http://www.thecampaign.org/>

This activist organization strives to keep American citizens informed about genetically engineered foods through the use of food labels. The site includes sample letters to send to legislators, news, education, forums, and more.

Colorado Genetic Engineering Action Network (COGEAN)
<http://www.foodlabeling.org/>

COGEAN is "a network of organizations in the state of Colorado working on the issue of genetically modified organisms." Site features include information, media, initiatives, and legislation. Although the content tends to be highly opinionated against genetic engineering, it cites some authoritative sources.

Food First: Genetic Engineering
<http://www.foodfirst.org/progs/global/ge/>

Food First is a nonprofit think tank that tackles world hunger and poverty. The site includes authoritative information about genetic engineering, including opinions, opposition, risks, and reports about genetically modified crops and foods.

GE Food Alert
<http://www.gefoodalert.org/>

GE (genetically engineered) Food Alert "is a coalition of seven organizations united in their commitment to testing and labeling genetically engineered food." Member organizations include the Public Interest Research Group, National Environmental Trust, Institute for

Agriculture and Trade Policy, Organic Consumers Association, Friends of the Earth, Center for Food Safety, and Pesticide Action Network of North America. Site features include news headlines, press releases, food alerts, action alerts, events, and links.

GE Free Food Guide
<http://www.greenpeace.org.nz/truefood/default.asp>

Greenpeace is the well-known activist organization that focuses on global environmental issues including "preventing the release of genetically engineered organisms into nature." This New Zealand site features information about genetically engineered foods, including this guide to foods that are not genetically engineered. The guide is browsable, searchable, and downloadable.

Genetic Engineering—Legal Briefs
<http://www.i-sis.org.uk/GE-legal.php>

The Institute of Science in Society (ISIS) is a nonprofit organization based in the United Kingdom. Although paid members have specific privileges (access to *Science in Society* and discounts on publications), the press releases about genetic engineering are freely available.

Genetic Engineering Action Network (GEAN)
<http://www.geaction.org/>

GEAN is "a diverse network of grassroots activists, national and community non-governmental organizations (NGOs), farmer and farm advocacy groups, academics and scientists who have come together to work on the myriad of issues surrounding biotechnology." Specifically, this organization focuses on choice, assessment, protection, and liability as related to genetic engineering. Site features include documents, news headlines, external links, and events.

Genetically Engineered Food
<http://www.organicconsumers.org/gelink.html>

Organic Consumers Association (OCA) is a nonprofit organization concerned with "food safety, industrial agriculture, genetic engineering, corporate accountability, and environmental sustainability." Site features include news stories, articles, alerts, and links to a variety of information including genetically engineered food, organic food, cloning, food poisoning, and more.

GeneWatch
<http://www.genewatch.org/>

GeneWatch is a United Kingdom–based nonprofit public interest group. Site information includes databases; publications; glossaries; press releases and other information about genetically modified crops, animals, and food; and biological weapons.

GM and Novel Foods
<http://www.food.gov.uk/gmfoods/>

The information on this site about genetically modified (GM) foods includes information about labeling, a news archive, links to European legislation, and more.

GreenpeaceUSA—Genetic Engineering
<http://www.greenpeaceusa.org/campaigns/intro?campaign_id= 503428>

Greenpeace is the well-known activist organization that focuses on global environmental issues including "preventing the release of genetically engineered organisms into nature."

Site information about genetic engineering includes an overview of the subject, the genetic engineering industry, food politics, sustainable alternatives (shopping lists, organic agriculture, etc.), news, publications, and multimedia.

Natural Food Commission
<http://www.naturallaw.org.nz/genetics/default.htm>

Organized by New Zealand's Natural Law Party, Natural Food Commission focuses on "the health hazards caused by the genetic engineering of food." Site features include fact sheets, papers, media releases, and a handbook on genetically engineered foods.

True Food Now!
<http://www.truefoodnow.org/>

The True Food Network is an activist organization that developed this site "as a part of Greenpeace's Genetic Engineering campaign in 2000." Now part of GEAN, the site focuses on supermarket activism; community events; and opinionated information about genetically engineered grain, biotechnology industries, and non–genetically engineered food brands.

HORMONES

Consumer Concerns About Hormones in Food
<http://envirocancer.cornell.edu/FactSheet/Diet/fs37.hormones.
  cfm>

Cornell University's Program on Breast Cancer and Environmental Risk Factors (BCERF) offers this understandable fact sheet about why hormones are used to raise meat and dairy products, a history of using hormones in the food industry, effects of hormone residues in food, increased risk of breast cancer from hormones in food, and recommendations to reduce exposure.

Growth Hormones
<http://www.hc-sc.gc.ca/vetdrugs-medsvet/growth_hormones_
  e.html>

Health Canada's Health Products and Food Branch explains the issues surrounding the administration of growth hormones to cattle and discusses research on the subject commissioned by the EU.

Hormones in Meat
<http://europa.eu.int/comm/food/food/chemicalsafety/contaminants/
  hormones/index_en.htm>

The European Commission "embodies and upholds the general interest of the [European] Union and is the driving force in the Union's institutional system." This site features special topics (avian influenza, bovine spongiform encephalopathy, GM food, GM feed, food labeling); news; resources (speeches, press releases, publications, events); and major site categories related to food and feed safety, animal health and welfare, plant health, and inspections. The information about hormones in meat introduces the topic, details European legislation, offers scientific opinion, and compares the European situation with the United States and Canada.

A Primer on Beef Hormones
<http://www.fas.usda.gov/itp/policy/hormone2.html>

The Foreign Agricultural Service (FAS) of the USDA provides this understandable explanation of hormones administered to beef cattle.

INSPECTIONS

Canadian Food Inspection Agency
<http://www.inspection.gc.ca/english/toce.shtml>

The Canadian Food Inspection Agency "delivers all federal inspection services related to food; animal health; and plant protection." Site features include hot topics, what's new, e-mail notification of recalls and alerts, and more.

Food Standards Agency
<http://www.food.gov.uk/>

Site features include information about bovine spongiform encephalopathy (mad cow disease); genetically modified and novel foods; labeling; interactive tools (games, quizzes, calculators); publications; news; and more. The information about inspections includes audit protocols, audit protocol checklists, and previsit questionnaires.

Sanitation Inspections of International Cruise Ships
<http://www2.cdc.gov/nceh/vsp/vspmain.asp>

The CDC publishes *Summary of Sanitation Inspections of International Cruise Ships,* also known as the "Green Sheet." This publication lists ships by name, inspection date, and score. Sanitation standards cover water; food preparation and holding; potential contamination of food; and general cleanliness, storage, and repair.

☑ USDA's Food Safety Inspection Service (FSIS)
<http://www.fsis.usda.gov/>

FSIS site features include news and events; fact sheets; and official forms. The site features browsable categories such as food safety education, science, regulations and policies, recalls, and emergency preparedness.

KITCHEN SAFETY

Can Your Kitchen Pass the Food Safety Test?
<http://www.fda.gov/fdac/features/895_kitchen.html>

The FDA is part of the Department of Health and Human Services of the U.S. government, and is responsible for "protecting consumers and promoting public health." This food safety test asks twelve questions about handling, defrosting, and cooking foods at home.

Food Safety in the Kitchen
<http://www.ces.ncsu.edu/depts/foodsci/agentinfo/hot/kitchen.html>

The North Carolina Cooperative Extension Service (CES), hosted at North Carolina State University (NCSU), collected and/or developed several documents about kitchen food safety. Topics include clean and safe kitchen, sponges and dishcloths, proper storage of food, hand washing, cutting boards, and other practices.

Food Safety in the Kitchen
<http://www.foodsafety.gov/~fsg/kitchen.html>

FoodSafety.gov is the U.S. government's gateway to information about food safety. The section on kitchen food safety includes links to local, state, and federal government sites.

Outdoor Food Preparation and Safety
<http://www.ext.vt.edu/pubs/nutrition/348-016/348-016.html>

Extension specialists at the Virginia Cooperative Extension at Virginia Polytechnic Institute and State University developed these tips about safely preparing food for outdoor consumption, especially in the summer months. Information includes basic safe preparation practices; food packing and storage; safe outdoor cooking techniques, including grilling; and safe cleanup.

LABELING, PACKAGING, AND STORAGE

The Campaign to Label Genetically Engineered Foods
<http://www.thecampaign.org/>

This organized effort by an activist organization focuses on action steps needed to keep American citizens informed about genetically engineered foods through the use of food labels. The site includes sample letters to send to legislators, news, education, forums, and more.

The Food Label
<http://www.fda.gov/opacom/backgrounders/foodlabel/newlabel. html>

The FDA is part of the Department of HHS of the U.S. government, and is responsible for "protecting consumers and promoting public health." The FDA site's information about new food labels discusses the Nutrition Labeling and Education Act of 1990 and defines terms included in the nutrition facts.

Food Labeling
<http://www.nal.usda.gov/fnic/etext/000027.html>

The Food and Nutrition Information Center (FNIC) is part of the USDA and the Agricultural Research Service (ARS), in partnership with the University of Maryland and Howard University. FNIC's subtopic on food labeling includes content and educational materials from the FDA, USDA, and other entities.

Food Labeling Overview
<http://vm.cfsan.fda.gov/label.html>

CFSAN is a department of the FDA, which in turn is part of the HHS. CFSAN's food labeling information includes an overview, general information, consumer nutrition, nutrient content, and industry regulations. The section on food labeling is enhanced with a search function and subject index to other topics.

Food Packaging Materials
<http://www.hc-sc.gc.ca/food-aliment/cs-ipc/chha-edpcs/e_food_ packaging.html>

Site information about food packaging materials includes Canadian regulations and guidelines.

Food Preservation and Safety
<http://hgic.clemson.edu/Site3.htm>

Clemson University's Home and Garden Information Center offers this information about preserving food. Site features include documents about canning, freezing, drying, and pickling foods; food handling and storage; foodborne illnesses; and emergencies. A typical document provides an overview of the subject and includes useful hints and tips.

Food Storage Tips
<http://www.hgof.ns.ca/index2.php?function=stor_tips>

Home Grown Organic Foods, a Canadian business, provides several practical tips for prolonging the life of foods and avoiding food contamination.

Labeling and Consumer Protection
<http://www.fsis.usda.gov/OPPDE/larc/Index.htm>

The FSIS of the USDA "is responsible for the safety and accurate labeling of meat, poultry, and egg products." Site information covers labeling procedures, policies, packaging materials, and more.

Recycled Plastics in Food Packaging
<http://vm.cfsan.fda.gov/~dms/opa-recy.html>

The CFSAN is a department of the FDA, which in turn is part of the HHS. The information about food packaging mentions FDA concerns that recycled plastics can contaminate foods.

Safe Food Storage Times and Temperatures
<http://www.hs.state.az.us/phs/oeh/fses/sfstt.htm>

The Office of Environmental Health in the Arizona Department of Health Services provides this information about safe lengths of time and temperatures to store food in the cupboard, refrigerator, and freezer. Individual charts list safe practices for categories of foods including meats; poultry; eggs and egg dishes; fish; milk, cream, and cheese; staples; mixed and packaged foods; canned and dried foods; other foods; vegetables; and fruits.

PESTICIDES

Consumer Information Center: Pesticides and Food
<http://www.pueblo.gsa.gov/cic_text/food/pesticides-andfood/food.html>

The Federal Citizen Information Center in Pueblo, Colorado, provides this information about pesticides and food from the Environmental Protection Agency (EPA), in consultation with the FDA and USDA. Areas of concern include vulnerability of infants and children to pesticide residue, reducing risk of exposure by eating a variety of foods, and consuming organically grown foods.

EPA and Food Security
<http://www.epa.gov/pesticides/factsheets/securty.htm>

The EPA "is responsible for a number of activities that contribute to food security within the United States, in areas such as food safety, water quality, and pesticide applicator training." Site information includes pesticide overview, evaluation, residue tolerances, safer pesticides, and integrated solutions such as agricultural, biotechnological, and international approaches.

Pesticide Residues
<http://www.cce.cornell.edu/food/fsarchives/050602/pesticide.html>

Cornell University's Cooperative Extension discusses pesticide residues in food, including whether organic foods have fewer pesticides, and links to external information on the subject.

Pesticides and Food: What You and Your Family Need to Know
<http://www.epa.gov/pesticides/food/>

The EPA site information includes government regulations, residue limits, protecting children, health practices, and health problems caused by pesticide residue.

Pesticides in Baby Food
<http://www.ewg.org/reports/Baby_food/baby_short.html>

The Environmental Working Group (EWG) is a "team of scientists, engineers, policy experts, lawyers and computer programmers" that examine "government data, legal documents, scientific studies" and other materials to discover potential environmental health threats and propose solutions. EWG's information about pesticide residues in baby food includes cancer, nervous system toxicity, reproductive and hormonal disorders, and other risks; food sampling and other testing methods; conclusions; and recommendations.

RESTAURANT EATING AND TAKEOUT FOODS

Eating Out Safety Tips
<http://www.healthatoz.com/healthatoz/Atoz/hl/sp/info/hfoodtips. jsp>

HealthAtoZ is a consumer health site developed by health professionals. Free access to all site information requires registration. The information about eating safely outside the home covers usefulness of spices and vinegar to kill some microorganisms and being particular about food that involves a lot of handling.

Food Safety at Home, School, and When Eating Out
<http://www.foodsafety.gov/~dms/cbook.html>

This gateway to U.S. government information about food safety includes news; safety alerts; content for children, teenagers, and educators; reporting of illnesses and complaints; foodborne pathogens; state and federal agencies; and food industry assistance. The information about eating at home, school, and elsewhere includes a coloring book for understanding basic principles such as keeping cold foods cold, hot foods hot, washing fruits and vegetables, and similar knowledge.

Salsa ingredients, Tabasco sauce, garlic, oregano, and many other spices have been shown to kill a variety of foodborne microorganisms.

Source: Eating Out Safely Tips <http://www.healthatoz.com/healthatoz/Atoz/h1/sp/info/hfoodtips.jsp>.

Restaurant Lookup
<http://www.pimahealth.org/chfs/ratings/index.asp>

The Pima County (Arizona) Health Department's Consumer Health and Food Safety site provides a searchable interface to restaurant food safety evaluation ratings. The database is searchable by restaurant name, alphabetic list, rating, and date. Although the data is limited to eating establishments in Pima County, details about critical violations are informative.

Safe Handling of Takeout Food
<http://www.fsis.usda.gov/OA/pubs/takeoutfoods.htm>

This consumer publication explains how to store, refrigerate, freeze, reheat, and discard takeout food.

Tips for Takeout
<http://www.crfa.ca/resourcecentre/foodsafety/owto/tipsfortake out.asp>

The Canadian Restaurant and Foodservice Association (CRFA) offers practical advice for storing and reheating takeout food leftovers for future consumption.

TERRORISM AND TAMPERING

Agroterrorism Information Page
<http://ianrhome.unl.edu/inthenews/agroterrorism.shtml>

The Institute of Agriculture and Natural Resources at the University of Nebraska at Lincoln has organized this information about agroterrorism. Resources include news articles, external sites, scientific articles, and bioterrorism information developed by the University of Nebraska.

Biosecurity and Bioterrorism: Biodefense Strategy, Practice, and Science
<http://www.medscape.com/viewpublication/907_index>

First released in 2003 by publisher Mary Ann Liebert and accessible through Medscape, this quarterly publication is "dedicated to bioscience, medical and public health response, infrastructure and institutions, international collaborations, agroterror/food safety, and citizen response and responsibility, as each of these issues relate to biodefense." Typical articles address issues such as aerosol dissemination of biological weapons, anthrax, and other topics. Medscape requires free registration (see Figure 6.2).

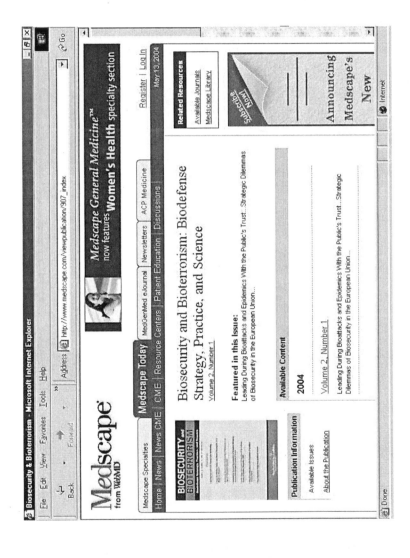

FIGURE 6.2. Biosecurity and Bioterrorism: Biodefense Strategy, Practice, and Science <http://www.medscape.com/viewpublications/907_index>

Biosecurity Awareness Guide
<http://www.afia.org/Biosecurity_Guide.html>

The American Feed Industry Association (AFIA) represents "manufacturers, ingredient suppliers, animal health companies, equipment manufacturers, large integrated livestock and poultry producers, and firms providing other goods and services to the commercial animal food industry." Site features include feed industry news, *Feedgram* newsletter which is free to members; and this *Biosecurity Awareness Guide* that "was produced with the participation of the Animal Health Institute (AHI); the Center of Veterinary Medicine (CVM)/Food and Drug Administration (FDA); and the National Renderers Association (NRA)."

Food Safety and Terrorism
<http://vm.cfsan.fda.gov/~dms/fsterr.html>

The Center for Food Safety and Applied Nutrition (CFSAN) is a department of the Food and Drug Administration (FDA), which in turn is part of the U.S. Department of Health and Human Services (HHS). CFSAN's information on terrorism includes an organized list of internal and external links about bioterrorism legislation; rapid tests of food contamination; registration of food facilities; and imported food shipments.

Food Tampering
<http://www.nzfsa.govt.nz/current-issues/food-tampering/>

The New Zealand Food Safety Authority (NZFSA) was established in 2002 as New Zealand's controlling authority for food import and export. NZFSA "protects and promotes public health and safety," and "facilitates access to markets for New Zealand food and food related products." NZFSA's information about food tampering covers media relations, fact sheets, reporting mechanisms, and links to external information.

Food Tampering
<http://www.cfsan.fda.gov/~dms/fstamper.html>

CFSAN's information includes tips for detecting food tampering at the supermarket and home, reporting food tampering, practicing healthy food preparation, and avoiding foodborne illnesses.

TOXINS

Aflatoxins
<http://www.oardc.ohio-state.edu/ohiofieldcropdisease/ Mycotoxins/mycopageaflatoxin.htm>

Ohio State University's information about moldy grains, myco-toxins, and feeding problems includes a section about crops and weather conditions that promote the development of aflatoxins, as well as allowable levels and health effects in animals.

Aflatoxins: Occurrence and Health Risks
<http://www.ansci.cornell.edu/plants/toxicagents/aflatoxin/ aflatoxin.html>

Cornell University's information about this by-product of storing crops includes factors that promote fungal growth, effects on animal and human health, analysis in foods and feeds, and more.

Mycotoxins and Food Supply
<http://www.fao.org/docrep/U3550t/u3550t0e.htm>

The FAO of the United Nations provides this information about molds in agricultural crops. Two scientists explain historical aspects, occurrence, health implications, economic implications, and prevention and control of mycotoxins in the food supply.

Natural Toxins
<http://www.food.gov.uk/news/newsarchive/2003/aug/ naturaltoxins>

This information explains the natural toxins in red kidney beans, green potatoes, moldy dough, and other foods.

Red Tide and Harmful Algal Blooms
<http://www.whoi.edu/redtide/>

Woods Hole Oceanographic Institution (WHOI) in Woods Hole, Massachusetts, is a nonprofit research facility that focuses on the study of marine science. WHOI shows a colorful graphic of how algae infiltrate the human food web through marine life.

WATER QUALITY

E. coli 0157:H7 in Drinking Water
<http://www.epa.gov/safewater/ecoli.html>

The EPA site explains the dangers of *E. coli* in drinking water, including health effects, symptoms, risk factors, methods to treat contaminated water supplies, and prevention.

Floods-Water Quality
<http://www.bt.cdc.gov/disasters/floods/water.asp>

The CDC explains the importance of preventing disease after a flood by maintaining water quality for drinking and cooking. Tips include using bottled water, boiling water, and/or treating water with chlorine or iodine.

Post Disaster Water Treatment
<http://www.redcross.org/services/disaster/0,1082,0_563_,00.html>

The American Red Cross (ARC) offers advice for treating water after a disaster. Site information describes how to filter, boil, and use chlorine bleach to make water potable, and kill "bacteria and parasites that cause diseases such as dysentery, cholera, typhoid, and hepatitis."

Chapter 7

Audiovisual and Multimedia Materials

The sites included in this section provide information about audiovisual and multimedia materials related to food safety and security. Some materials are intended for the layperson, and others are better suited for health professionals and scientists. Consult Chapter 8 on education and conferences for related information. The symbol ☑ is used to denote major resources with authoritative and original content.

Dirty Little Secrets
<http://www.fda.gov/opacom/catalog/videos/secrets/secrets.html>

The Food and Drug Administration (FDA) is part of the Department of Health and Human Services (HHS) of the U.S. government, and is responsible for "protecting consumers and promoting public health." The site offers a ten-second clip from its eight-and-one-half minute videotape about kitchen food safety, which is for sale through the FDA.

Food Protection Program
<http://www.metrokc.gov/health/foodsfty/videos/>

King County in Seattle, Washington, maintains a site that includes information about food safety, including streaming video clips (available in seven languages) about the importance of handwashing and preventing food contamination. RealOne Player is needed to view these clips, but the site provides a link to download the free software.

Food Safety Music
<http://foodsafe.ucdavis.edu/music.html>

The University of California at Davis maintains a FoodSafe program that focuses on consumer and industry information related to food safety. The site includes synthesized music and parodied lyrics developed by a food toxicologist. The site also includes PowerPoint slides related to these song parodies.

Food Safety Slides and Video on the Web
<http://peaches.nal.usda.gov/foodborne/fbindex/slides_videos.asp>

The Foodborne Illness Education Information Center, maintained by the National Agricultural Library (NAL), developed this list of food safety multimedia that is available through government and nongovernment Web sites.

Food Safety Videos
<http://peaches.nal.usda.gov/foodborne/fbidb/videos.asp>

The NAL describes the food safety videotapes that can be borrowed from their document delivery branch, or through a local library that borrows the material(s) from NAL.

FSIS Image Library
<http://www.fsis.usda.gov/OA/pubs/image_library/index.htm>

The Food Safety Inspection Service (FSIS) is "the public health agency in the U.S. Department of Agriculture [USDA] responsible for ensuring that the nation's commercial supply of meat, poultry, and egg products is safe, wholesome, and correctly labeled and packaged." Their image library includes artwork, graphics, and links to clip art related to food safety.

Government Food Safety Video Library
<http://www.foodsafety.gov/~fsg/vlibrary.html>

FoodSafety.gov is the U.S. government's gateway to information about food safety. The site's food safety video library includes video clips that can be viewed over the Internet using RealPlayer software.

☑ International Food Safety Icons
<http://www.foodprotection.org/>

The International Society for Food Protection licenses the use of international icons to depict food safety practices, including hand washing, cooking temperatures, avoiding food handling when ill, avoiding cross-contamination of foods, washing surfaces and utensils, wearing gloves to handle ready-to-eat foods, cooling temperatures, keeping cold foods cold, keeping hot foods hot, and others. These images can be freely downloaded for educational and noncommercial purposes (see Figure 7.1).

Hand washing: Wash hands with soap and warm running water.

Wash, Rinse, Sanitize: Food contact surfaces and utensils must be properly washed, rinsed, and sanitized.

Refrigeration/Cold Holding: Cold foods must be held at 41°F (5°C) or below.

Temperature Danger Zone: Do not allow foods to stay in the temperature danger zone. Keep cold foods cold and hot foods hot.

Hot Holding: Hot foods must be held at 140°F (60°C) or above.

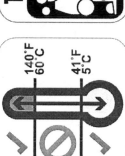

Cross Contamination: Do not cross-contaminate between raw and ready-to-eat or cooked foods.

Cooking: Thoroughly cook foods to appropriate temperatures.

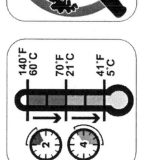

Cooling: Hot foods must be cooled from 140 to 70°F (60 to 21°C) within 2 hours and from 70 to 41°F (21 to 5°C) within an additional 4 hours.

FIGURE 7.1. International Food Safety Icons. (*Source:* Copyright © International Association for Food Protection.)

☑ Public Health Image Library
<http://phil.cdc.gov/Phil/default.asp>

The U.S. Centers for Disease Control and Prevention's (CDC) Public Health Image Library (PHIL) includes illustrations, animations, and other multimedia related to public health. Examples include gram-stained specimens; photographs of persons, animals, and places; scanning electron micrographs; photomicrographs; and video clips. Many of these images and files are in the public domain, meaning that they can be freely used without obtaining permission.

Chapter 8

Education and Conferences

The sites included in this section cover many aspects of food safety and security education. Meetings and conferences are excellent sources of information for health professionals and scientists interested in these fields. Generally, these events are geared toward industry experts rather than laypersons, but research and policy decisions presented and discussed in these settings have implications for food safety and security. The symbol ☑ is used to denote major sites related to this category.

☑Canadian Partnership for Consumer Food Safety Education
<http://www.canfightbac.org/english/indexe.shtml>

The Canadian Partnership for Consumer Food Safety Education is "a national association of public and private organizations committed to educating Canadians about the ease and importance of food safety in the home." Site information includes safe handling tips, FightBAC images and logos, an interactive game, free teaching materials that can be requested by e-mail or telephone, fact sheets, and public service announcements. Site content is also presented in the French language.

FNIC Resource Lists
<http://www.nal.usda.gov/fnic/pubs_and_db.html>

Food and Nutrition Information Center (FNIC) is part of the USDA and the Agriculture Research Service (ARS), in partnership with the University of Maryland and Howard University. The FNIC site includes a list of annotated database links. The databases related to food safety include AGRIS/CARIS/FAO Online, produced by the Food and Agricultural Organization (FAO) of the United Nations; Agriculture Network Information Center (AgNIC) Agriculture Database; FNIC databases; International Bibliographic Information on Dietary Supplements (IBIDS) database; AGRICultural OnLine Access (AGRICOLA); Dietary Exposure Potential Model; Medline Plus; U.S. Department of Agriculture Bibliography Databases; Food Preservation Database, produced by Pennsylvania State University; and other resources (see Figure 8.1).

Food Safety Education
<http://www.fsis.usda.gov/education.index.asp>

The Food Safety Inspection Service (FSIS) is "the public health agency in the U.S. Department of Agriculture responsible for ensuring that the nation's commercial supply of meat, poultry, and egg products is safe, wholesome, and correctly labeled and packaged." The FSIS site organizes food safety education and consumer information, including *The Food Safety Educator,* a newsletter that is available free on this site and "distributed to nearly 10,000 educators throughout the country including public health offices, extension educators, industry, and consumer groups."

Food Safety Education
<http://www.foodsafety.gov/~dms/fsebac.html>

FoodSafety.gov is the U.S. government's gateway to information about food safety. Site features include information about the four basic food safety steps (clean, separate, cook, and chill); what's new in government food safety; and food safety brochures and information sheets in the English, Chinese, and Spanish languages.

Food Safety Summit
<http://www.foodsafetysummit.com/>

Food Safety Summit is a yearly conference that focuses on food safety and quality assurance for food manufacturers, processors, restaurant chains, and supermarkets. Typical topics of concern include bioterrorism, bovine spongiform encephalopathy, listeria, federal regulations, and more.

International Conference on Emerging Infectious Diseases (ICEID)
<http://www.iceid.org/default.asp>

Sponsored by the U.S. Centers for Disease Control and Prevention (CDC), American Society for Microbiology (ASM), the World Health

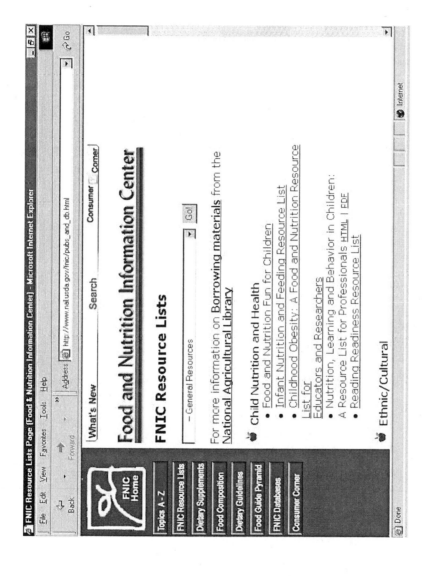

FIGURE 8.1. FNIC Resource Lists <http://www.nal.usda.gov/fnic/pubs_and_db.html>

Organization (WHO), and others, the first ICEID was held in 1998. Typical meetings convene

> public health professionals to encourage the exchange of scientific and public health information on global emerging infectious disease issues. . . . Major topics include current work on surveillance, epidemiology, research, communication and training, bioterrorism, and preventions and control of emerging infectious diseases, both in the United States and abroad.

Typically, the conference site includes abstracts, video clips of papers presented, and other useful information.

Chapter 9

Interactive Tools

This section's sites include databases, calculators, quizzes, and tools related to food safety and security and aspects of consumer health. The symbol ☑ is used to denote major resources with authoritative and original content.

A to Z Comprehensive List of Terms
<http://www.cfsan.fda.gov/~dms/a2z-term.html>

The Center for Food Safety and Applied Nutrition (CFSAN) is a department of the Food and Drug Administration (FDA), which in turn is part of the U.S. Department of Health and Human Services (HHS). This alphabetical list of terms ranges from acidification to mycotoxins to zoonosis. Typical entries include understandable definitions, illustrations, and implications for food safety.

Agency for Toxic Substances and Disease Registry
<http://www.atsdr.cdc.gov/HAC/PHA/>

The CDC's Agency for Toxic Substances and Disease Registry (ATSDR) focuses on protecting Americans from exposure to toxic substances and situations by identifying and cleaning up hazardous waste sites. ATSDR's public health assessments refer to evaluations of hazardous exposures. These assessments are searchable by keyword and browsable by state. The relevance to food safety includes information about dioxins, toxic substances in fish and shellfish, and other information.

☑ AGRICOLA
<http://agricola.nal.usda.gov/>

The National Agricultural Library (NAL) in Beltsville, Maryland, provides access to AGRICOLA, an article citation database that indexes journal articles, book chapters, short reports, and reprints. NAL's

Web-based catalog of their library holdings can be searched at the same time as AGRICOLA.

AgView.com
<http://www.agview.com/>

Developed in 1996, this search interface locates data, information, and resources related to agriculture. Search results are annotated. Subject categories include agriculture, crop production, food science, biotechnology, organic farming, weather, and more.

Ask the Nutrition Expert
<http://www.cce.cornell.edu/food/>

Cornell University's Cooperative Extension Food and Nutrition focuses on "research-based information for consumers, educators and other professionals in the areas of food, nutrition and health, and food safety." Nutrition professionals are invited to ask a nutrition expert questions related to a featured topic. Past topics have included dietary reference intakes, lead exposure, plant estrogens, and antioxidants.

Can Your Kitchen Pass the Food Safety Test?
<http://www.fda.gov/fdac/features/895_kitchen.html>

The FDA developed this simple test to determine whether a home kitchen follows safe food practices. The information was originally published in a 1995 issue of *FDA Consumer* but has been updated several times since then.

☑ Combined Health Information Database (CHID)
<http://chid.nih.gov/>

CHID is a bibliographic database that combines the consumer health efforts of U.S. government agencies. Typical search results yield bibliographic citations to consumer health publications available from a variety of sources.

Compendium of Food Additive Specifications
<http://apps3.fao.org/jecfa/additive_specs/foodad-q.jsp>

The Joint FAO/WHO Expert Committee on Food Additives (JECFA) developed this database of food additives (other than flavors). FAO is the Food and Agriculture Organization of the World Health Organization (WHO). The content is available in Arabic,

French, Spanish, and English languages. Entries can be searched by substance name, International Numbering System (INS) number, Chemical Abstracts Service (CAS) number, functional use group (antioxidants, emulsifiers, stabilizers, etc.), purity group (cadmium, lead, arsenic, etc.), and food additives designated as tentative. Typical database entries include substance name, synonyms, description, chemical name(s), chemical formula, structural formula, characteristics, method of assay, and more.

FNIC Databases
<http://www.nal.usda.gov/fnic/databases.html>

The NAL's Food and Nutrition Information Center (FNIC) is part of the U.S. Department of Agriculture (USDA) and the Agricultural Research Service (ARS), in partnership with the University of Maryland and Howard University. This search interface allows aggregated searching of various databases including food safety, hazard analysis and critical control point (HACCP) training materials, FSRIO (Food Safety Research Information Office), IBIDS (International Bibliographic Information on Dietary Supplements), and other nutrition resources (see Figure 9.1).

Food Consumer.com
<http://foodconsumer.com/forum/index.php>

The FoodConsumer.com site features a moderated discussion forum devoted to food laws and regulations; chemicals in food; food preparation; food, diet, and nutrition; and related topics.

Food Irradiation Index
<http://www.hc-sc.gc.ca/food-aliment/fpi-ipa/e_irradiation_index. html>

Health Canada is Canada's federal health agency that works with provincial and territorial governments "to develop health policy, enforce health regulations, promote disease prevention and enhance healthy living for all Canadians." The site indexes information about industry requests to irradiate foods including shrimp, mangoes, ground beef, and poultry. Index information summarizes a specific request and its evaluation.

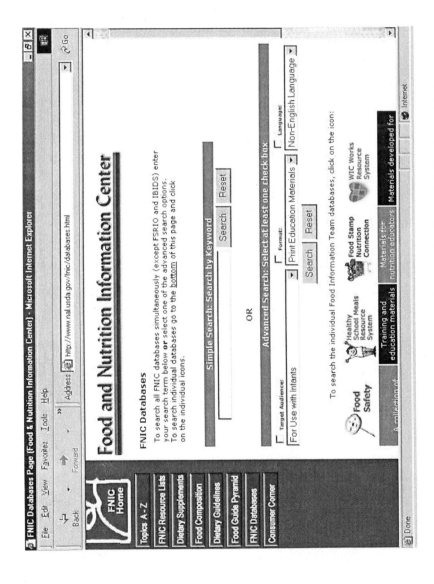

FIGURE 9.1. FNIC Databases <http://www.nal.usda.gov/fnic/databases.html>

Food Navigator
<http://www.foodnavigator.com/>

This Europe-based resource provides breaking news about foods and beverages, covering a "large range of ingredients and additives including: nutraceuticals, vitamins and minerals, flavors, enzymes, colors, emulsifiers, fibers, preservatives and hydrocolloids."

Food Safety Lessons
<http://www.exnet.iastate.edu/foodsafety/Lesson/glossary.html>

The Food Safety Project at the Iowa State University developed these food safety lessons and a glossary "to provide consumers and future consumers with the tools they need to help minimize their risk from harmful pathogens in the food supply."

Food Safety Quizzes
<http://www.dupagehealth.org/safefood/quiz/storage_quiz.asp>

The DuPage County Health Department in Illinois developed this series of quizzes related to food preparation, food storage, and food temperature knowledge.

☑ Food Safety Risk Analysis Databases
<http://www.foodriskclearinghouse.umd.edu/databases.cfm>

The FDA and the University of Maryland's Food Safety Risk Analysis Clearinghouse organized this group of freely accessible and fee-based databases. The resources are organized into broad categories, including agriculture, biotechnology, chemical hazards, commodities, economic, food intake, food temperature, geospatial, health, international, natural resources, nutrition, papers and documents, pathogens, social science, and U.S. statistics.

Foodlink—Fun and Games
<http://www.foodlink.org.uk/testyourself.asp>

The United Kingdom's Food and Drink Federation organizes and maintains the Foodlink site. The interactive food safety games include Calamity Kitchen, Quiz Time, Word Search, and Spot the Difference.

☑ FSRIO Database
<http://peaches.nal.usda.gov/fsrio/fsrioform.asp>

FSRIO developed this database of research related to food safety. Search results can be restricted to category (contaminants and contamination, diseases and poisonings, food and food products, food handling and processing, on-farm food safety, etc.) and federal or state agency. Typical search results include abstracts, contact information for principal investigator(s), and links to research project reports.

HACCP Database
<http://vm.cfsan.fda.gov/~lrd/haccp.html>

The information about HACCP includes an overview, news, highlights, specific information about seafood and juice, and HACCP educational activities. Federal and state governments use HACCP as process controls for the food service industry.

HazDat
<http://www.atsdr.cdc.gov/hazdat.html>

HazDat is the ATSDR's scientific database about the effects of hazardous substances on human populations, site characteristics, activities and site events, contaminants found, contaminant media and maximum concentration levels, impact on population, community health concerns, ATSDR public health threat categorization, ATSDR recommendations, environmental fate of hazardous substances, exposure routes, and physical hazards at the site/event.

Hazmat Safety
<http://hazmat.dot.gov/>

The U.S. Department of Transportation (DOT) helps promote national safety and prevent dangerous situations caused by the transportation of hazardous materials. Hazmat is a contracted word formed from *haz*ardous *mat*erials. Site features include regulations, documentation, news, discussion, training, and more related to chemical safety and emergency response systems. Site content specifically related to food safety and security includes genetically modified microorganisms, packaging standards, and infectious substances.

Interactive Kitchen
\<http://www.homefoodsafety.org>

HomeFoodSafety.org combines the efforts of the American Dietetic Association (ADA) and the ConAgra Foods Foundation. This nine-question quiz assesses knowledge of safe household handling, preparation, and storage of foods.

Mutagens in Cooked Foods Database
\<http://www-bio.llnl.gov/mutagens/html/db.intro.text.html>

Lawrence Livermore National Laboratory (LLNL) hosts this database about food mutagens. Typical entries include information about specific foods, their preparation (fried, grilled, heated), mutagens, risks, and references.

National Food Safety Database (NFSD)
\<http://foodsafety.ifas.ufl.edu/indexNFSDB.htm>

The University of Florida Institute of Food and Agricultural Sciences (IFAS) Extension provides food safety information for consumers, educators, and industry professionals alike. Site features include daily news, with some information in the Spanish language. Topics of interest to consumers include holiday food safety; labeling; food handling; microwave safety; canning, drying, and freezing; hotlines; and more. Resources for educators include HACCP information, preparing for food/water shortages, and links to federal and state agencies. Topics of interest to industry professionals include food production, processing, research, testing, and government agencies.

Population Survey Atlas of Exposures
\<http://www.cdc.gov/foodnet/surveys/Pop_surv.htm>

The U.S. Centers for Disease Control and Prevention (CDC) focuses on protecting health and safety by monitoring and preventing disease. FoodNet is the CDC's surveillance network for foodborne diseases. *Population Survey Atlas of Exposures* is a searchable interface to exposure frequency by site for different types of food (fruits and vegetables; meat, poultry, and seafood; dairy and eggs; herbs, water, etc.) and factors (children in day care, meal locations, consumer knowledge) for the years 1996 to 2000.

Summer Food Safety Quiz
<http://www.gnb.ca/0053/foodsafety/BarbecueSafety-e.asp>

Canada's New Brunswick Department of Health and Wellness developed this quiz about summer food safety. The quiz is available in English and French languages.

USDA Meat and Poultry Hotline
<http://www.fsis.usda.gov/Food_safety_education/USDA_Meat_ &_Poultry_Hotline/index.asp>

The Food Safety Inspection Service (FSIS) of the USDA provides e-mail and toll-free telephone access to information about "safe storage, handling, and preparation of meat, poultry, and egg products." The site also provides links to selected publications, activity reports, and frequently asked questions.

Chapter 10

Organizations and Societies

The sites included in this section cover organizations and societies that focus on food safety and security. Consult Chapter 12 for specific documents developed by some of these organizations, societies, and other entities.

Agency for Toxic Substances and Disease Registry
<http://www.atsdr.cdc.gov/HAC/PHA/>

The U.S. Centers for Disease Control and Prevention's (CDC) Agency for Toxic Substances and Disease Registry (ATSDR) focuses on protecting Americans from exposure to toxic substances and situations by identifying and cleaning up hazardous waste sites. ATSDR's public health assessments refer to evaluations of hazardous exposures. These assessments are searchable by keyword and browsable by state. The relevance to food safety includes information about dioxins, toxic substances in fish and shellfish, and other information.

American Dietetic Association (ADA)
<http://www.eatright.org/Public/>

The ADA's membership includes food scientists, dietitians, and nutritionists. The organization focuses on five major areas of concern: obesity, aging, dietary supplements, safe and nutritious food supply, and genetics. The site includes position papers on food and nutrition topics, nutrition fact sheets, information about dietary public policy, career information, conferences, and more (see Figure 10.1).

American Society for Microbiology (ASM)
<http://www.asm.org/>

Established in 1899, ASM's "mission is to advance microbiological sciences through the pursuit of scientific knowledge and dissemi-

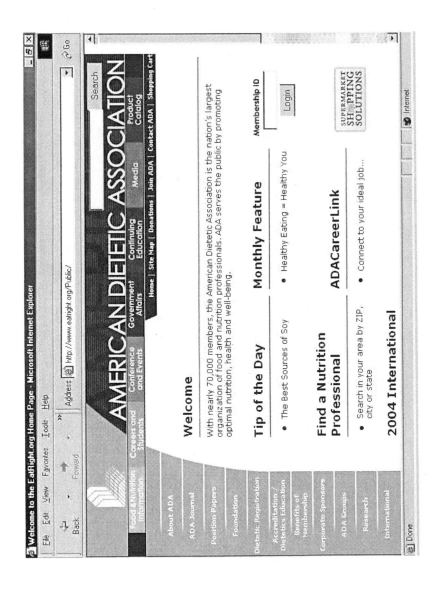

FIGURE 10.1. American Dietetic Association (ADA) <http://www.eatright.org/Public/>

nation of the results of fundamental and applied research." The ASM site includes information about membership, publications, education, policy, and other information.

Center for Consumer Research (CCR)
<http://ccr.ucdavis.edu/>

CCR is part of the University of California at Davis, and its information on food safety focuses on "consumer attitudes toward food safety and quality." Biotechnology information includes general information, news, hazards, benefits, safety, regulations, and message board. Food irradiation information features general information, what's new, myths, safety issues, regulation, consumer acceptance, and message boards.

Center for Science in the Public Interest (CSPI)
<http://www.cspinet.org/>

CSPI is a Washington, DC-based nonprofit organization devoted to consumer advocacy in several areas of public health. Areas of focus include biotechnology, antibiotic resistance, food safety, nutrition and health, alcohol policy, and integrity in science. Site features include newsroom, reports, finding elected officials by ZIP code, press releases, and more.

Community Food Security Coalition (CFSC)
<http://www.foodsecurity.org/>

CFSC is headquartered in Venice, California. This nonprofit organization focuses on creating "a system of growing, manufacturing, processing, making available, and selling food that is regionally based and grounded in the principles of justice, democracy, and sustainability." Site features include what's new, programs, events, publications, and views. Members can access the coalition newspaper and receive discounts on publications.

Ecological and Toxicological Association of Dyes and Organic Pigments Manufacturers (ETAD)
<http://www.etad.com/>

ETAD is an international organization of more than forty-five dye manufacturers in fifteen countries. "Formed in 1974 to represent the interests of these industries on matters relating to health and environment," the organization strives to reduce adverse health effects of dyes and pigments. Site features include code of ethics, objectives, annual reports, literature database, upcoming events, and external links. Some site features are restricted to organization members.

Environmental Health Center
<http://www.nsc.org/ehc.htm>

The National Safety Council's Environmental Health Center concerns itself with air quality, environmental health for children, climate change, disaster recovery, hazardous chemicals, radiation, radioactive wastes, solid wastes, water, and coasts. Information related to food safety includes unintentional poisonings, hazardous chemicals, and recalls.

Food Standards Agency
<http://www.food.gov.uk/>

The United Kingdom's Food Standards Agency was established by Parliament in 2000 "to protect the public's health and consumer interests in relation to food." Site features include news, bovine spongiform encephalopathy (mad cow disease), food labeling, genetically modified foods, research, and interactive tools.

GRACE Factory Farm Project (GFFP)
<http://www.factoryfarm.org/>

GFFP "works to create a sustainable food production system which is healthful and humane, economically viable and environmentally sound." In this case, sustainable refers to organic farming, and there is a bias against factory farms—otherwise described as concentrated animal feeding operations (CAFO). Site features include resources by topic and region; solutions (quick facts, reports, organizations); news and news archive; activism; and more. Highlights include a guide to salmon shopping; *Eat Well Guide;* an animated video clip about Meatrix, a defender of family farms; photographs of factory farms; and more.

Health Physics Society
<http://www.hps.org/>

Health physics is the field of knowledge related to ionizing radiation hazards. The Health Physics Society is a professional organization that focuses on educational and research activities related to this subject, particularly as related to environmental and occupational risks. Site features include news; radiation facts; position papers, including ones on food irradiation; and an "ask the experts" section.

IFDA Online
<http://www.ifdaonline.org/>

IFDA is the International Foodservice Distributors Association, an organization that focuses on research, education, and communication for persons employed by food service industries. Food safety resources include IFDA publications about hazard analysis and critical control points (HACCP), role of bar codes in food safety, and more; links to articles from external sites; and industry links.

International Association for Food Protection (IAFP)
<http://www.foodprotection.org/>

IAFP is a nonprofit organization of food safety professionals, including educators, government employees, industry professionals, and equipment manufacturers. The IAFP site features tables of contents and abstracts for *Journal of Food Protection, Food Protection Trends,* and other publications; news; workshops; audiovisual library; conferences; and more.

International Food Information Council (IFIC)
<http://ific.org/food/>

IFIC is a Washington, DC-based organization whose mission is "to communicate science-based information on food safety and nutrition to health and nutrition professionals, educators, journalists, government officials and others providing information to consumers." IFIC activities are "supported primarily by the broad-based food, beverage and agricultural industries." Site features include news, newsletters, fact sheets, posters, and more.

Joint Institute for Food Safety and Applied Nutrition (JIFSAN)
<http://www.jifsan.umd.edu/whatjifsan.htm>

JIFSAN is a joint effort between the University of Maryland and the U.S. Food and Drug Administration (FDA). JIFSAN's mission is to "provide the scientific basis for ensuring a safe, wholesome food supply as well as provide the infrastructure for contributions to national food safety programs and international food standards." Site features include a good agricultural practices (GAPs) manual, information about educational programs and internships, and upcoming events.

National Coalition for Food Safe Schools (NCFSS)
<http://www.foodsafeschools.org/>

NCFSS is a group of public and private organizations with representatives from national organizations, associations, and government agencies. The coalition works together to reduce "foodborne illness in the U.S. by improving food safety in schools." Intended for students, teachers, and parents alike, the site serves as a gateway to food safety information by organizing and annotating links to external sites.

National Food Processors Association (NFPA)
<http://www.nfpa-food.org/>

NFPA is an organization of persons who "process and package fruits, vegetables, meat, fish, and specialty food and beverage products." Site features include information about public policy, food science, food security, publications, and workshops. The site restricts some content to NFPA members, but nonmembers will appreciate the information about listeria, food labeling, compliance with bioterrorism regulations, and food security.

New Zealand Food Safety Authority (NZFSA)
<http://www.nzfsa.govt.nz/>

NZFSA was established in 2002 as New Zealand's controlling authority for food import and export. NZFSA "protects and promotes public health and safety," and "facilitates access to markets for New Zealand food and food related products." Site features include food safety topics for consumers, policy and law, industry, and science and technology. Consumer information includes food safety topics (additives, antibiotic resistance, irradiation), food-handling tips, marine contamination warnings, and reporting food safety concerns.

Organic Consumers Association (OCA)
<http://www.organicconsumers.org/>

OCA is a nonprofit organization concerned with "food safety, industrial agriculture, genetic engineering, corporate accountability, and environmental sustainability." Site features include information and news about genetically engineered foods, organic foods, food irradiation, food contamination, and risks from specific foods (beef, venison, seafood).

Pan American Health Organization (PAHO)
<http://www.paho.org/>

PAHO serves as the regional office for the Americas (North America, Latin America, Central America, the Caribbean, South America) for the World Health Organization (WHO) and functions as part of the United Nations. Site features include data (health indicators, country health profiles, health trends); topics (communicable diseases, nutrition and food protection, zoonoses); and resources (announcements, publications, technical documents) (see Figure 10.2).

Woods Hole Oceanographic Institution (WHOI)
<http://www.whoi.edu/>

WHOI is located in Woods Hole, Massachusetts, and "is dedicated to research and higher education at the frontiers of ocean science." Site features include general information, educational outreach, research, and publications of interest to researchers, educators, and the public.

World Health Organization (WHO)
<http://www.who.int/>

WHO focuses on many aspects of international health, including prevention of diseases and promotion of "physical, mental and social well-being." The site is organized into major categories of interest: countries, health topics, publications, and research tools. Content is featured in English, French, and Spanish languages.

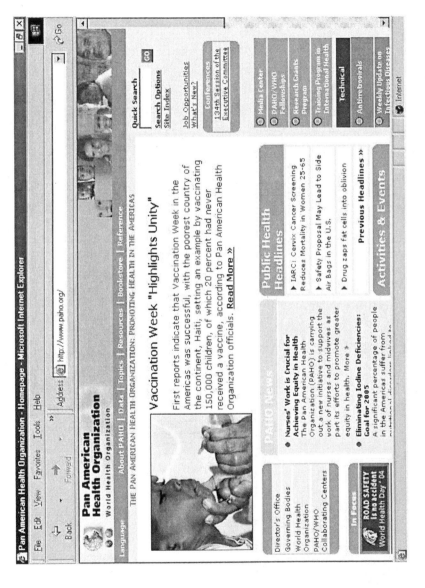

FIGURE 10.2. Pan American Health Organization (PAHO) <http://www.paho.org/>

Chapter 11

Legislation, Standards, and Regulations

The sites included in this chapter cover many aspects of food legislation, standards, and regulations. The symbol ☑ is used to denote major resources with authoritative and original content.

Australian Food Safety Standards
<http://www.nswfitc.com.au/a/1864.html>

Australia's New South Wales (NSW) Food Industry Training Council is

> an industry training advisory body (ITAB) funded to . . . provide independent, representative advice to Government on the training needs and priorities of each industry sector across food and beverage processing, tobacco processing and pharmaceutical manufacturing; advise industry on the implementation of quality training systems, including available funding, resources and best practice; and act as industry's advocate and voice on policy and related issues.

Site features include news and standards.

Bioterrorism Act of 2002
<http://www.fda.gov/oc/bioterrorism/bioact.html>

Several sections of the Bioterrorism Preparedness and Response Act of 2002 pertain to food safety and security, such as food supply protection and food adulteration. The U.S. Food and Drug Administration (FDA) site features the act itself in HTML and PDF formats, plans for implementing the legislation, and other useful bioterrorism information.

☑ European Union Food Safety Legislation
<http://www.eurunion.org/legislat/Foodstuffs/FoodSafetyLeg.htm>

The Delegation to the European Union (EU) in the United States site covers legislation related to animal feed, health and welfare of food-producing animals, food contaminants, residues, dietetic foods and supplements, food additives and flavors, genetically modified (GM) foods, irradiated food products, and packaging.

Food Packaging Materials
<http://www.hc-sc.gc.ca/food-aliment/cs-ipc/chha-edpcs/e_food
_packaging.html>

Health Canada is Canada's federal health agency that works with provincial and territorial governments "to develop health policy, enforce health regulations, promote disease prevention and enhance healthy living for all Canadians." Site information about food packaging materials includes regulations and guidelines.

Food Safety/Legislation
<http://www.afia.org/Government_Affairs/FDA_CVM_Issues/
Food_Safety.html>

The American Feed Industry Association (AFIA) represents "manufacturers, ingredient suppliers, animal health companies, equipment manufacturers, large integrated livestock and poultry producers, and firms providing other goods and services to the commercial animal food industry." The Government Affairs section of the AFIA site summarizes issues such as feed labeling, antibiotic use in food animals, dioxin, ingredient definitions, public perception of genetically engineered (GE) products, and a single federal food agency. Typical summaries include issue, importance, status, AFIA position, opposing positions, AFIA action, and future outlook.

Food Safety Links—Legislation and Regulation
<http://peaches.nal.usda.gov/foodborne/fbindex/Regulations.asp>

The U.S. Department of Agriculture (USDA) and FDA combine their efforts in this foodborne illness education information center. This list of food safety legislation and regulation includes links to government and nongovernment sites.

Food Safety Standards
<http://www.foodstandards.gov.au/mediareleasespublications/publications/foodsafetystandardscostsandbenefits/>

Food Standards Australia New Zealand "protects the health and safety of the people in Australia and New Zealand by maintaining a safe food supply." The site includes an unofficial, consolidated version of the Australia New Zealand Food Standards Code. Parts of this code cover general food standards, food product standards, food safety standards, primary production standards, labeling, additives, contaminants, residues, microbiological limits, processing requirements, standards for different types of foods, preliminary provisions, transitional standards, and amendments.

Protecting the Food Supply
<http://www.cfsan.fda.gov/~dms/fsbtact5.html>

The Center for Food Safety and Applied Nutrition (CFSAN) is a department of the FDA, which in turn is part of the U.S. Department of Health and Human Services (HHS). The information about protecting the food supply refers to the Bioterrorism Preparedness and Response Act of 2002 and explains new regulations proposed by the FDA especially related to registering food facilities, advance notice of imported food shipments, and detection of potentially health-threatening foods. This document is available in English, Spanish, French, Bulgarian, Croatian, Czech, Hungarian, Polish, Romanian, Russian, Serbian, Slovak, Slovene, and Ukrainian languages.

Select Agent Program
<http://www.cdc.gov/od/sap/>

The U.S. Centers for Disease Control and Prevention (CDC) site provides information about the select agent program, which requires the "registration of facilities including government agencies, universities, research institutions, and commercial entities," that possess biological agents and toxins. Information includes regulations, application materials, list of select agents, and more.

Chapter 12

Publications

The sites included in this section feature full-text, freely accessible documents related to food safety and security. The symbol ☑ is used to denote major resources with authoritative and original content.

Basics for Handling Food Safely
<http://www.fsis.usda.gov/oa/pubs/facts_basics.htm>

The Food Safety Inspection Service (FSIS) is "the public health agency in the U.S. Department of Agriculture [USDA] responsible for ensuring that the nation's commercial supply of meat, poultry, and egg products is safe, wholesome, and correctly labeled and packaged." This basic information covers safe steps for handling, cooking, and storing foods; details cold storage (refrigerator and freezer) for specific foods; and more.

Biosecurity and the Food Supply
<http://www.fsis.usda.gov/oa/background/biosecurity.htm>

This fact sheet explains the roles of the USDA, FSIS, and other government entities; various biosecurity activities conducted by the Office of Homeland Security and Food Biosecurity Action Team (F-BAT); and other collaborative efforts.

Biosecurity Awareness Guide
<http://www.afia.org/Biosecurity_Guide.html>

The American Feed Industry Association (AFIA) represents "manufacturers, ingredient suppliers, animal health companies, equipment manufacturers, large integrated livestock and poultry producers, and firms providing other goods and services to the commercial animal food industry." This four-page guide is available in PDF format, along with a related PowerPoint presentation in the English and

Spanish languages. Guide information includes adulteration of feeds with biological, chemical, radiological, and physical ingredients; security of buildings and grounds; ingredient integrity; product distribution; product recall; employee selection and training; and emergency response.

☑Canadian Partnership for Consumer Food Safety Education Fact Sheets
<http://www.canfightbac.org/english/ccentre/factsheets/factsheetse.shtml>

The Canadian Partnership for Consumer Food Safety Education is "a national association of public and private organizations committed to educating Canadians about the ease and importance of food safety in the home." The fact sheets cover topics such as bacterial foodborne illness in Canada; causes of foodborne illness; bacteria; and separate fact sheets on the importance of separating, cooking, cleaning, and chilling foods safely.

☑ *CDC Fact Book 2000/2001*
<http://www.cdc.gov/maso/factbook/main.htm>

Although not geared specifically toward food safety, the Centers for Disease Control and Prevention's *CDC Fact Book 2000/2001* includes information about plans to improve health and prevent foodborne diseases; hepatitis A, B, and C; and other illnesses.

☑ *Diagnosis and Management of Foodborne Illnesses: A Primer for Physicians and Other Health Care Professionals*
<http://www.ama-assn.org/ama/pub/category/3629.html>

The American Medical Association (AMA), American Nurses Association (ANA), CDC, Center for Food Safety and Applied Nutrition (CFSAN) of the Food and Drug Administration (FDA), and the Food Safety and Inspection Service (FSIS) of the USDA developed this handbook for health care professionals. The publication, now in its second edition, is downloadable free from this site or by e-mail request. Chapters cover foodborne illnesses caused by bacterial, viral, parasitic, and noninfectious agents. Clinical scenarios feature patients with antibiotic-resistant salmonellosis, botulism poisoning, congenital toxoplasmosis, *E. coli* infections, hepatitis A, listeria monocytogenes, norovirus infection, and unexplained illnesses. This publication also includes the *FightBAC* handout, suggested food safety resources, a reading list, and continuing medical education questions.

☑ *Disease Outbreak News*
<http://www.who.int/csr/don/en/>

The Communicable Disease Surveillance and Response (CSR) department of the World Health Organization (WHO) publishes *Disease Outbreak News* as a medium for disseminating information to contain known risks, respond to unexpected health threats, and improve preparedness throughout the world. Archived issues from 1996 to the present can be sorted by disease, year, and country (see Figure 12.1).

EPA Publications
<http://www.epa.gov/epahome/publications2.htm>

The U.S. Environmental Protection Agency (EPA) is "responsible for a number of activities that contribute to food security within the United States, in areas such as food safety, water quality, and pesticide applicator training." EPA publications related to food safety and security include documents developed by the Office of Water, Office of Pollution Prevention and Toxics, Pollution Prevention Information Clearinghouse, and Office of Science and Technology.

Eurosurveillance Weekly
<http://www.eurosurveillance.org/ew/index-02.asp>

The European Commission funds the weekly publication of *Eurosurveillance Weekly,* "an electronic bulletin for epidemic alerts, updates, and responses." Issues date back to the first volume, published in 1997 (see Figure 12.2).

Food and Agriculture Factsheets
<http://www-tc.iaea.org/tcweb/publications/factsheets/>

The International Atomic Energy Association (IAEA) is an agency within the United Nations (UN) that focuses on the promotion of "safe, secure and peaceful nuclear technologies." IAEA's publications includes several booklets about food irradiation.

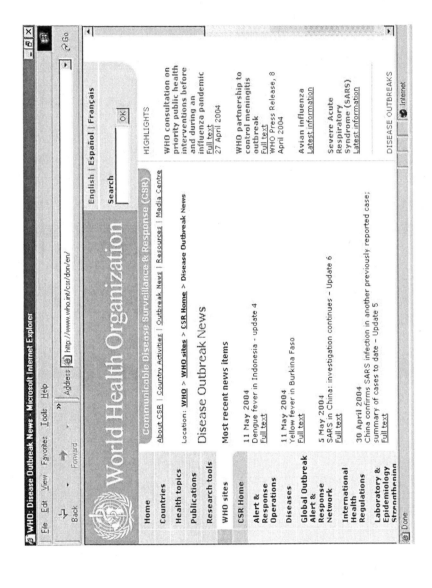

FIGURE 12.1. *Disease Outbreak News* <http://www.who.int/csr/don/en/>

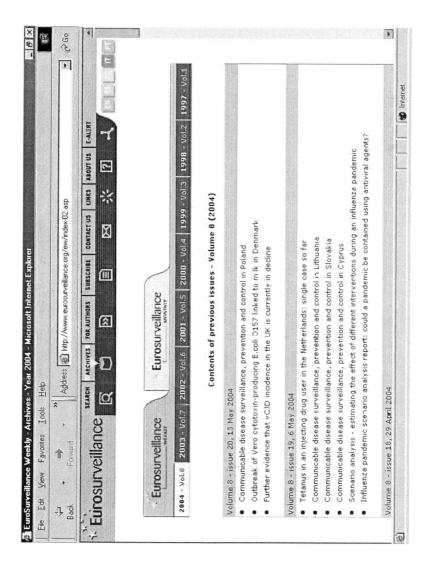

FIGURE 12.2. *Eurosurveillance Weekly* <http://www.eurosurveillance.org/ew/index-02.asp>

Food Irradiation
<http://hps.org/hpspublications/papers.html>

The Health Physics Society is a professional organization that focuses on educational and research activities related to ionizing radiation. This two-page position statement details the dangers of food irradiation.

Food Preservation and Safety
<http://hgic.clemson.edu/Site3.htm>

Clemson University's Home and Garden Information Center offers this information about preserving food. Site features include documents about canning, freezing, drying, and pickling foods; food handling and storage; foodborne illnesses; and emergencies. A typical document provides an overview of the subject and includes useful hints and tips.

Food Safety and Food Security: What Consumers Need to Know
<http://www.fsis.usda.gov/oa/topics/foodsec_cons.htm>

The FSIS of the USDA published this document in November 2003. The fifteen-page publication includes tips for preparing foods safely and avoiding food contamination. It is available in English and Spanish languages.

☑ *Food Safety and Hygiene*
<http://www.foodscience.afisc.csiro.au/fshlist.htm>

Food Science Australia is a joint venture between Commonwealth Scientific and Industrial Research Organisation (CSIRO) and the Australian Food Safety Centre of Excellence. This site provides a searchable interface to *Food Safety and Hygiene* issues from March 1995 to the present. Typical issues include articles about preventing meatborne contamination, genetically modified (GM) foods, international spread of foodborne illness, microbiology of organic vegetables, and other topics.

☑ *Food Safety Educator*
<http://www.fsis.usda.gov/news_and_events/food_safety_educator/ index.asp>

FSIS publishes *The Food Safety Educator,* a free quarterly newsletter that covers topics such as pathogens, safe food handling, food irradiation, safe water, and more.

Foodborne Illness
**<http://www.tdh.state.tx.us/ideas/foodborne_illness/charts_
pamphlets/>**

The Texas Department of State Health Services site features several useful charts and posters related to foodborne illness. Topics range from preventing specific foodborne illnesses, foodborne illness symptom onset represented in both chart and poster form, and etiology of foodborne illness by food and season.

☑ *Foodborne Pathogenic Microorganisms and Natural Toxins*
  Handbook
<http://vm.cfsan.fda.gov/~mow/intro.html>

Published by the Center for Food Safety and Applied Nutrition, a department of the FDA, this handbook includes useful information about foodborne pathogenic microorganisms (bacteria, viruses, and parasites) and natural toxins that can cause diseases in humans. Typical entries feature the name of the organism or group of organisms; nature of disease (acute symptoms, onset time, infective dose, duration of symptoms, and cause of disease); diagnosis of human illness; associated foods; relative frequency of disease; reported cases; complications; susceptible populations, foods analysis, selected outbreaks, education, and other resources. Also called the *Bad Bug Book*.

**FSIS Safety and Security Guidelines for the Transportation and
  Distribution of Meat, Poultry and Egg Products**
<http://www.fsis.usda.gov/oa/topics/transportguide.pdf>

This forty-page document outlines safety and security guidelines for the transportation and distribution of specific food products.

GE Free Food Guide
<http://www.greenpeace.org.nz/truefood/default.asp>

Greenpeace is the well-known activist organization that focuses on global environmental issues including "preventing the release of genetically engineered organisms into nature." Its New Zealand site features information about genetically engineered (GE) foods, including this guide to foods that are not genetically engineered. The guide is browsable, searchable, and downloadable.

Guide to Minimize Microbial Food Safety Hazards for Fresh Fruits and Vegetables
<http://www.cfsan.fda.gov/~dms/prodguid.html>

CFSAN is a department of the FDA, which in turn is part of the U.S. Department of Health and Human Services (HHS). Information covers definitions, water, manure, worker health and hygiene, sanitary facilities, field sanitation, packing facility sanitation, transportation, and more. This guide is available in the English, French, Portuguese, Spanish, and Arabic languages.

National Agricultural Library (NAL) Publications and Databases
<http://www.nal.usda.gov/pubs_dbs/pubs.htm>

NAL has several publications on its site that relate to food safety and security. Examples include annual reports, NAL staff directories, land grant university directories, newsletters, and more.

Natural Carcinogens and Anticarcinogens in America's Food
<http://www.acsh.org/publications/booklets/nature.html>

The American Council on Science and Health (ACSH) is a nonprofit organization that concerns itself "with issues related to food, nutrition, chemicals, pharmaceuticals, lifestyle, the environment and health." The site includes a forty-one-page document about cancer-causing and cancer-fighting substances in foods. Written by physicians associated with reputable institutions, the publication defines and explains carcinogens and mutagens, aflatoxins and other mold toxins, toxins in common foods, cancer-fighting substances in foods, and hazards from natural and synthetic substances.

☑ NutriNet
<http://fscn.che.umn.edu/outreach/faculty_outreach/nutrinet.html>

The University of Minnesota developed NutriNet as a means to disseminate information about food science and nutrition. The site's food safety newsletter is published monthly. Typical issues include food safety insights, consumer-friendly features that can be clipped and saved, and resources (fact sheets, Web sites, conferences).

Nutrition and Food Safety Publications
<http://lancaster.unl.edu/food/pubs.htm>

The University of Nebraska Cooperative Extension in Lancaster County developed these brief nutrition and food safety "publications," edited and updated by a registered dietitian. Food safety topics include shigella, clostridium botulinum, salmonella, and trichinella spiralis.

Glossary

Some of the terms used throughout this guide may be unfamiliar to the layperson. These definitions and information about acronyms were derived from several excellent sources:

- **HyperDictionary**
  <http://www.hyperdictionary.com/>
- **Medical Dictionary**
  <http://www.medical-dictionary.com/>
- **MedlinePlus Merriam-Webster Medical Dictionary**
  <http://www.nlm.nih.gov/medlineplus/mplusdictionary.html>
- **MedTerms.com Medical Dictionary**
  <http://www.medterms.com/>
- **OneLook Dictionary Search**
  <http://www.onelook.com/>

acronyms: Words formed from the first letter or letters of words in a name or phrase, such as CDC for Centers for Disease Control and Prevention.

acrylamide: Harmful substance that forms when carbohydrates are cooked at high temperatures.

adulteration: The act of making something impure or inferior by adding foreign or impure ingredients.

aflatoxins: Carcinogenic MYCOTOXINS that may result when agricultural crops are stored.

aggregated: Collected or combined. Related to consumer health information, this term is used to indicate grouped resources that can be searched with a common interface.

agroterrorism: The use of biological or chemical agents to harm or produce disease in agricultural crops. Agriculture + terrorism = agroterrorism.

algae: Simple, microscopic plants that live in water.

allergy: High sensitivity to irritants such as pollens, foods, or MICRO-ORGANISMS, resulting in physical reactions.

amebiasis: Infection or disease caused by amebas, protozoan organisms that can cause diarrheal illness. Amebiasis is often caused by the entamoeba histolytica organism.

anaphylaxis: Immediate and severe allergic reaction, sometimes resulting in shock and death.

anthocyanins: Red or blue plant pigments believed to have health benefits.

antibodies: Proteins created by the body's immune system to fight infections or harmful substances.

antigens: Substances used to stimulate an immune response.

antimicrobials: Substances that kill or reduce infectious agents.

antioxidants: Substances that slow the aging process and prevent disease; examples include vitamin E, beta-carotene, and BHT (butylated hydroxytoluene). Cereal manufacturers, for example, use BHT to coat cereal boxes to help keep the products fresh.

antitoxins: Substances capable of neutralizing or reducing the actions of TOXINS.

aquaculture: Farming of fish in artificial tanks, ponds, or net enclosures.

assay: Analysis of a substance to determine presence, absence, or quantity of one or more ingredients or parts.

authoritative: As related to Web site content, accurate and reliable.

avian: Related to birds.

bacteria: Single-celled organisms that live independently or feed parasitically off other organisms. Bacteria can be harmless or harmful.

bibliographic: As related to informational databases, citations that include author, title, source, and abstract, elements that are needed for finding full-text information in print or electronic publications.

biologicals: Globulins, sera, vaccines, ANTITOXINS, or ANTIGENS used to prevent or treat diseases.

biosensors: Equipment that uses BIOLOGICALS to detect chemicals in a substance.

biotechnology: The use of MICROORGANISMS or BIOLOGICALS in specific industrial or manufacturing processes; the application of engineering and technology principles to the life sciences.

blue sheet: *Summary of Health Information for International Travel* as developed by the U.S. Centers for Disease Control and Prevention (CDC).

bovine: Related to cattle.

bovine spongiform encephalopathy: A neurodegenerative disease in cattle caused by a PRION that has been transmitted by feed infected with animal tissue; also called mad cow disease.

carcinogens: Cancer-causing agents or substances.

carnivores: Animals that feed on foods derived from animals or plants that feed on insects.

carotenoids: Yellow or red PHYTOCHEMICALS thought to have health benefits. Examples include apricots, carrots, tomatoes, peaches, and corn.

channels: As related to Web sites, specific subsections of content. One site may have several channels devoted to specific topics or audiences.

chemoprophylaxis: The use of drugs to prevent disease. *See also* PROPHYLAXIS.

CJD: *See* CREUTZFELDT-JAKOB DISEASE.

clinical trials: Series of treatments used to evaluate the effectiveness of specific medications or medical procedures.

communicable diseases: Diseases that are contagious or easily transmitted from one organism to another.

congeners: Substances, plants, or animals that are related.

contamination: Accidental or intentional pollution with infectious, hazardous, toxic, radioactive, or foreign matter.

contraindications: Conditions or diseases that preclude using a particular treatment or procedure.

Creutzfeldt-Jakob disease: A rare, progressive, and fatal neurodegenerative disease that is marked by the development of SPONGIFORM ENCEPHALOPATHY (porous brain tissue), premature DEMENTIA in middle age, and gradual loss of muscular coordination. *See also* PRIONS.

cross-contamination: To soil or infect by close contact or association.

degenerative: Deteriorating or showing loss of function.

dementia: Deterioration of intellectual abilities such as memory, concentration, and judgment, due to an organic disease or a disorder of the brain.

diagnosis: Methods used to identify characteristics, signs, or symptoms of a disease, condition, or ailment that distinguishes the disease from other diseases.

dioxins: Toxic substances that result from various manufacturing processes, especially waste management, paper production, and pesticides. Also called POLYCHLORINATED DIBENZO-*P*-DIOXINS or PCDDs.

disease outbreaks: Sudden increases or eruptions of disease activity.

disease surveillance: Close observation and tracking of disease activity by public health authorities. *See also* SURVEILLANCE.

disease transmission: Transfer of disease to another organism.

disease vectors: Organisms such as mosquitoes or ticks that carry disease-causing MICROORGANISMS from one host (person or animal) to another person or animal.

disinfection: The process of cleaning that destroys harmful organisms or prevents the spread of disease.

DNA: Deoxyribonucleic acid, the material inside cells that controls cell functions and controls the inheritance of traits and characteristics.

encephalopathy: A DEGENERATIVE disease of the brain.

endemic: Diseases common or peculiar to a specific geographic region. Compare this term with ENDEMIC.

endotoxins: Poisons released by specific bacterial organisms.

enteric: Related to or within the intestinal tract.

enterococci: Normally harmless bacteria that flourish in the intestinal tract.

environmental justice: Meaningful participation in decisions related to health and environment, regardless of socioeconomic level.

epidemic: The outbreak and spread of disease in a geographic area. Compare this term with ENDEMIC.

epizootic: A disease that affects many organisms at the same time within a specific region or geographic area.

Escherichia coli: Known more familiarly as *E. coli,* this organism normally exists within the gastrointestinal system, but dangerous forms are responsible for serious blood or intestinal diseases.

etiology: The study of the causes of disease.

fauna: Animals living in a specific region. The term *flora and fauna* is used to denote the plants and animals of a geographic area.

fermentation: The process by which a yeast is used to convert sugar into alcohol.

flavonoids: PHYTOCHEMICALS that are thought to have health benefits. Examples include the ANTHOCYANINS.

flora: Plants growing in a specific region. The term *flora and fauna* is used to denote the plants and animals of a geographic area.

food contaminants: Agents or substances that taint, pollute, or poison foods or beverages. Examples include BACTERIA, VIRUSES, chemicals, and PARASITES.

food contamination: The intentional or unintentional tainting, polluting, or poisoning of foods or beverages.

food poisoning: Injury, illness, or death caused by contaminated foods or beverages.

foodborne: Carried, transmitted, or caused by foods or beverages.

fortify: To enrich food by adding ingredients such as vitamins and minerals.

fungi: Plural of fungus; plant organisms that include yeasts, molds, mildews, mosses, algaes, smuts, and mushrooms.

furans: Abbreviation of POLYCHLORINATED DIBENZOFURANS, toxic compounds resulting from various manufacturing processes including incineration and paper production. Furans can accumulate in biological tissues and have been found to contaminate eggs and meat products. Also called PCDFs.

gastroenteritis: Stomach or intestinal upset including nausea, vomiting, and diarrhea, often caused by FOOD POISONING.

genetic: Relating to the science of genetics, or the mechanisms by which an organism's traits are inherited.

genetic engineering: The alteration of GENETIC MATERIAL in a living organism.

genetic material: This is a general term that covers organic human materials such as hair, saliva, and blood, as well as DNA sequences that have been extracted from human cells.

genetic mutation: Alteration or change of GENETIC MATERIAL.

genetically modified organisms: Plants or animals that have had their genetic makeup altered or modified in ways that could not be accomplished through normal reproduction. Foods that have been genetically modified (GM) are sometimes called NOVEL FOODS.

genome: An organism's GENETIC MATERIAL.

globulins: Proteins found in blood, milk, muscle, and plant seeds.

gray literature: Publications that are not widely available through normal channels.

green sheet: The U.S. Centers for Disease Control and Prevention (CDC) publishes *Summary of Sanitation Inspections of International Cruise Ships,* also known as the "Green Sheet."

HACCP: Hazard Analysis/Critical Control Point, a term used by several organizations, including the United States Department of Agriculture (USDA), to determine hazardous substances and situations.

hazard analysis: Identification of substance properties or situations that can result in harm or loss of life.

heavy metals: Substances such as cadmium, lead, nickel, copper, iron, and zinc, especially known to have adverse health effects.

hemorrhagic: Characterized by sudden and/or excessive loss of blood.

hepatitis: Inflammation of the liver caused by infection or toxins.

herbicides: Chemicals used to kill or slow the growth of plants, especially weeds.

herbivores: Humans or animals that feed chiefly on grasses and other plants.

hormones: Substances that affect growth or METABOLISM.

host: An organism on which another organism (PARASITE) feeds or lives.

immune response: A physical response to an ANTIGEN by specific ANTIBODIES.

immunity: The capacity to resist disease.

immunization: A procedure that introduces specific ANTIGENS to induce an IMMUNE RESPONSE, in an effort to improve the ability to resist infection and to render IMMUNITY for a specific disease. This term is used interchangeably with VACCINATION.

immunocompromised: Lack of a normal immune response, as a result of disease, malnutrition, or immunosuppressive therapy.

immunosuppression: Suppression of the IMMUNE RESPONSE by disease, medications, or radiation.

inorganic: Not composed of organic material. Compare this term with ORGANIC.

insecticides: Chemicals used to kill insects. *See also* PESTICIDES.

invisible Web: A term that refers to Internet content not easily accessible by normal search engines because of the way the information is organized.

ionizing radiation: High-energy radiation such as X-rays. Ionizing radiation can be used to sterilize, preserve, or prolong the life of foods.

irradiation: The use of ionizing radiation to sterilize or preserve food.

lethal dose: The amount of a substance known to cause death.

lycopene: Red pigments in tomatoes, papayas, watermelons, and other ripe fruits that are believed to have health benefits.

MEDLINE: A bibliographic database of the published biomedical literature, developed by the U.S. National Library of Medicine (NLM).

MedlinePlus: A consumer health resource developed by the National Library of Medicine that provides extensive information about more than 650 diseases and conditions. Also includes lists of hospitals and physicians, medical encyclopedia, medical dictionary, information about prescription and nonprescription drugs, and health information from the media.

metabolism: Chemical processes within a living organism.

microorganisms: Tiny life forms such as BACTERIA or PROTOZOA that can be harmful or beneficial.

mirex: An organochlorine INSECTICIDE.

morbidity: Relative incidence or rate of disease.

mortality: Relative incidence or rate of death.

mutagens: Substances thought to increase GENETIC MUTATION.

mycobacteria: A type of bacteria, some of which cause tuberculosis and leprosy.

mycotic: Infections or diseases caused by a fungus or fungi.

mycotoxins: Toxins produced by a fungus or fungi.

neurotoxin: A poisonous substance that affects the nervous system.

notifiable diseases: Diseases required by law to be reported to public health authorities.

novel foods: GENETICALLY MODIFIED foods or foods that are new to a geographical region.

nutraceuticals: Foods or food supplements thought to have beneficial health effects.

nutrient: Substances such as trace elements that promote growth.

occurrence: Instances of a specific disease appearing in a population.

omnivores: Persons or animals that eat both animal and plant foods.

organic: Derived from living organisms. Compare this term with INORGANIC.

outbreak: Sudden increase or eruption of disease.

PAHs: POLYCYCLIC AROMATIC HYDROCARBONS are chemicals that are produced by many industrial processes, but also result when foods are grilled.

parasites: Organisms that feed or live on another organism, often called the HOST.

pasteurization: The process of heating or sterilizing foods or beverages to prevent spoilage, FERMENTATION, or disease-causing organisms. Named after chemist Louis Pasteur (1822-1895), who perfected the process.

pathogenic: An agent or substance than can cause disease. *See also* PATHOGENS.

pathogens: Organisms, agents, or substances that cause a specific disease. *See also* PATHOGENIC.

PCBs: POLYCHLORINATED BIPHENYLS, environmental pollutants that accumulate in animal tissue; known to cause a variety of PATHOGENIC and TERATOGENIC effects.

PCDDs: POLYCHLORINATED DIBENZO-*P*-DIOXINS, toxic compounds commonly formed during waste management, incineration, paper production, and pesticides manufacture. Also called DIOXINS.

PCDFs: POLYCHLORINATED DIBENZOFURANS, toxic compounds resulting from various industrial processes including waste management, incineration, paper production, and pesticides manufacture. These compounds can accumulate in biological tissues and have been found to contaminate eggs and meat products. Also called FURANS.

peer reviewed: Professional evaluation of a colleague's work. This term usually refers to the evaluation of articles for publication or grants for funding.

pesticides: Chemicals used to kill pests, especially insects, worms, and other organisms.

phytochemicals: Plant substances such as FLAVONOIDS or CAROT-ENOIDS which are thought to have health benefits.

polychlorinated biphenyls: Environmental pollutants that accumulate in animal tissue, known to cause a variety of PATHOGENIC and TERATOGENIC effects. Also called PCBs.

polychlorinated dibenzofurans: Toxic compounds resulting from various industrial processes including waste management, incineration, paper production, and pesticides manufacture. These compounds can accumulate in biological tissues and have been found to contaminate eggs and meat products. Also called PCDFs or FURANS.

polychlorinated dibenzo-*p*-dioxins: Toxic compounds resulting from various industrial processes including waste management, incineration, paper production, and pesticides manufacture. Also called PCDDs or DIOXINS.

polycyclic aromatic hydrocarbons: Chemicals that are produced by many industrial processes, but also result when foods are grilled. Also called PAHs.

portals: As related to Web sites, gateways to resources.

potable: Describes a liquid, especially water, that is safe to drink.

primer: An introductory textbook or guide.

prions: Protein particles similar to viruses thought to cause CREUTZ-FELDT-JAKOB DISEASE and other degenerative diseases of the nervous system. The term is derived from the combination of the words proteinaceous + infectious = prion.

prophylaxis: Protective or preventive measures taken to prevent specific diseases.

proteinaceous: Related to protein or proteins. *See* PROTEINS.

proteins: Complex organic substances essential for growth and repair of tissues in humans and animals. Proteins are commonly found in meat, eggs, dairy products, and some plant foods.

protozoa: Single-celled organisms including amebas, ciliates, flagellates, and sporozoans.

public announcement: A type of written communication used by the U.S. State Department to inform American citizens about terrorism threats and other short-term conditions that affect safety and security while traveling to or living in other countries.

pulsed-field gel electrophoresis (PFGE): A process used to determine genetic TOXICANTS in fish and other animal life.

quarantine: Isolation of an individual or animal to prevent the spread of disease.

reagent: A chemical that is used to produce a specific chemical reaction.

residue: In this context, the amount of a toxic substance that remains on surfaces, in the soil, or in plant or animal tissues.

search directory: A Web tool that organizes Internet resources by subject and/or file type (documents, images, newsgroup messages).

search engine: A Web tool that finds Internet resources based on keywords or phrases typed into a search interface.

sera: The plural of serum, a constituent part of blood that often contains ANTIBODIES.

seroconversion: The body's development of ANTIBODIES in reaction to ANTIGENS.

species: Category of related organisms that are capable of breeding with one another.

spongiform: With a soft, porous, and spongelike texture. *See also* BOVINE SPONGIFORM ENCEPHALOPATHY.

subset: A part of a larger collection.

surveillance: The close observation of persons or situations. *See also* DISEASE SURVEILLANCE.

survivalists: Individuals who focus on and prepare for physical survival (food, water, shelter, weapons) during natural or man-made disasters, wars, or political or social upheavals.

sustainability: Practices used to maintain a healthy environment, particularly recycling.

teratogenic: Causing birth defects. *See also* TERATOGENS. Compare this term with MUTAGENS.

teratogens: Drugs or substances thought to cause birth defects. Compare this term with MUTAGENS.

think tanks: Organizations that study, research, and report on significant societal issues.

tick-borne: Diseases carried or transmitted by ticks.

tissues: As related to the body, a collection of similar cells that act together and perform specific functions.

toxicants: Poisons or poisonous substances.

toxicology: The study of poisons and their effects on the body.

toxins: Harmful substances that can cause disease.

transgenic: Relating to organisms that have been altered through the transfer of GENETIC MATERIAL from another species or breed.

transmission: Transfer of a disease from one person or animal to another.

vaccination: Introduction of harmless or killed bacterial or viral organisms to render IMMUNITY for a specific disease; a procedure that introduces specific ANTIGENS to induce an IMMUNE RESPONSE, thereby improving the ability to resist infection. This term can be used interchangeably with IMMUNIZATION.

vaccines: Preparations of MICROORGANISMS used to stimulate an IMMUNE RESPONSE to prevent future infection with similar microorganisms.

vectors: Ticks, mosquitoes, or other organisms that transfer disease-carrying MICROORGANISMS from one host to another.

viruses: Tiny parasites that feed on plants, animals, and bacteria, often causing diseases.

viticulture: The cultivation of grapes.

Web browser: A software program such as Netscape or Internet Explorer used to view content (text, images) on the World Wide Web.

white paper: An AUTHORITATIVE report or position paper on an issue of concern or controversy.

yellow book: Also known as *Health Information for International Travel*, published by the U.S. Centers for Disease Control and Prevention (CDC).

zoonosis: TRANSMISSION of disease from animals to humans. The plural form of this term is zoonoses.

Bibliography

DeWaal, Caroline Smith. "Safe Food from a Consumer Perspective." *Food Control* 14(2003): 75-79.

Engel, Karl-Heinz, Takeoka, Gary R., and Teranishi, Roy. *Genetically Modified Foods.* Washington, DC: American Chemical Society, 1995.

Redman, Nina E. *Food Safety: A Reference Handbook.* Santa Barbara, CA: ABC-CLIO, 2000.

Satin, Morton. *Food Alert! The Ultimate Sourcebook for Food Safety.* New York: Facts on File, 1999.

Wilcock, Anne, Pun, Maria, Khanona, Joseph, and Aung, May. "Consumer Attitudes, Knowledge and Behaviour: A Review of Food Safety Issues." *Trends in Food Science & Technology* 15(February 2004): 56-66.

Index

Page numbers followed by the letter "f" indicate figures; those followed by the letter "t" indicate tables.

Order a copy of this book with this form or online at:
http://www.haworthpress.com/store/product.asp?sku=5373

INTERNET GUIDE TO FOOD SAFETY AND SECURITY

_____in hardbound at $29.95 (ISBN-13: 978-0-7890-2631-6; ISBN-10: 0-7890-2631-7)

_____in softbound at $19.95 (ISBN-13: 978-0-7890-2632-3; ISBN-10: 0-7890-2632-5)

Or order online and use special offer code HEC25 in the shopping cart.

COST OF BOOKS_____	☐ **BILL ME LATER:** (Bill-me option is good on US/Canada/Mexico orders only; not good to jobbers, wholesalers, or subscription agencies.)
POSTAGE & HANDLING_____ *(US: $4.00 for first book & $1.50 for each additional book)* *(Outside US: $5.00 for first book & $2.00 for each additional book)*	☐ Check here if billing address is different from shipping address and attach purchase order and billing address information. Signature_____
SUBTOTAL_____	☐ **PAYMENT ENCLOSED: $**_____
IN CANADA: ADD 7% GST_____	☐ **PLEASE CHARGE TO MY CREDIT CARD.**
STATE TAX_____ *(NJ, NY, OH, MN, CA, IL, IN, PA, & SD residents, add appropriate local sales tax)*	☐ Visa ☐ MasterCard ☐ AmEx ☐ Discover ☐ Diner's Club ☐ Eurocard ☐ JCB Account # _____
FINAL TOTAL_____ *(If paying in Canadian funds, convert using the current exchange rate, UNESCO coupons welcome)*	Exp. Date_____ Signature_____

Prices in US dollars and subject to change without notice.

NAME_____

INSTITUTION_____

ADDRESS_____

CITY_____

STATE/ZIP_____

COUNTRY_____ COUNTY (NY residents only)_____

TEL_____ FAX_____

E-MAIL_____

May we use your e-mail address for confirmations and other types of information? ☐ Yes ☐ No We appreciate receiving your e-mail address and fax number. Haworth would like to e-mail or fax special discount offers to you, as a preferred customer. **We will never share, rent, or exchange your e-mail address or fax number.** We regard such actions as an invasion of your privacy.

Order From Your Local Bookstore or Directly From
The Haworth Press, Inc.
10 Alice Street, Binghamton, New York 13904-1580 • USA
TELEPHONE: 1-800-HAWORTH (1-800-429-6784) / Outside US/Canada: (607) 722-5857
FAX: 1-800-895-0582 / Outside US/Canada: (607) 771-0012
E-mail to: orders@haworthpress.com

For orders outside US and Canada, you may wish to order through your local sales representative, distributor, or bookseller.
For information, see http://haworthpress.com/distributors

(Discounts are available for individual orders in US and Canada only, not booksellers/distributors.)
PLEASE PHOTOCOPY THIS FORM FOR YOUR PERSONAL USE.
http://www.HaworthPress.com   BOF04